AF342291

Man and the Steam Engine

Two *Fowler* steam engines hauling a
plough across a field.

Man and the
Steam Engine

G. Watkins and R. A. Buchanan

PRIORY PRESS LIMITED

SBN 85078 149 3

Copyright © 1975 by G. Watkins and R. A. Buchanan
First published in 1975 by
Priory Press Ltd
101 Grays Inn Rd., London WC1

Text set in 12/14 pt Photon Baskerville, printed by photolithograph
and bound in Great Britain at The Pitman Press, Bath

Contents

1: *Introduction*

The modern story of the steam engine goes back to the seventeenth century, but in some respects it was anticipated in the Ancient World. In the later Roman Empire, a Greek, Hero of Alexandria, invented several ingenious steam-powered devices. These included doors which opened and closed seemingly of their own accord, and a spinning sphere called an "aeolipile" propelled by small steam jets. These and other imaginative inventions of the period could well have been put to use, but they were never regarded as more than amusing toys, and were quickly forgotten.

Why did a society with the best minds of Greek and Roman civilization show so little interest in the practical possibilities of steam power? Although we cannot fully answer this question, we can find out which social conditions are likely to lead to technological innovations. One possibility is that the inventions of Hero and others in the field of steam power were not used because nobody needed them. There were various reasons for this. The Roman Empire had an economic system which was more or less stagnant, a political system which placed more emphasis on stability than on change, and a social system which was heavily dependent upon slaves for menial tasks. Intelligent men regarded any sort of practical activity concerned with industrial production and craftsmanship to be unworthy of their attention. All this added up to an unwillingness to put money and skill into the development of inventions which could increase productivity, and a lack of interest in innovation except as a hobby for the aristocracy.

Opposite. Hero demonstrating his steam-powered "aeolipile" about A.D.100. The Ancient Greeks and Romans were not very interested in the practical application of such inventions.

Right. Galileo Galilei, whose experimental approach to science heralded the start of the scientific revolution of the seventeenth century.

From this knowledge of the Ancient World we can tell which social conditions are necessary for successful inventions in general and for the steam engine in particular. Firstly, if a good idea is ever to become a practical invention, it must fulfil a need—and this need must be important enough for the society to want to solve it. Secondly, an invention can never become a commercial success unless enough capital, skilled labour and materials are made available to produce it cheaply. Thirdly, the society in question (or at least an influential section of that society) must be prepared to tolerate inventors and to experiment with their ideas.

For a new idea to be a success, especially when it involves a complete technology like steam power, it is not sufficient for an individual inventor just to produce ideas—however brilliant they may be. Unless the invention triggers off a strong sense of social need and a willingness to devote social resources to its development, it is difficult to see how anything can happen. Such a combination of circumstances only existed occasionally in the Ancient World, and by the later centuries of the Roman Empire it had disappeared completely. Thus the ingenious devices of Hero never got beyond the stage of being models and toys. The successful development of steam technology could only be realized in an adventurous intellectual climate, and in a society which knew its needs and was prepared to use its skills to solve them.

The set of conditions necessary for the emergence of steam technology occurred in Western Europe as a result of the profound changes which took place between the fifteenth and seventeenth centuries. These changes resulted in the pattern of nation states with which we are still familiar in Europe; in the expansion of European civilization by the voyages of discovery; in the growth of trade; in the forward-looking intellectual attitude springing from the Renaissance and from the "scientific revolution." Science of a sort is as old as civilization, and certainly the Greeks had advanced scientific ideas, especially in mathematics and geometry. But the addition of an experimental method and—perhaps more significantly—the notion that science was a practical way of increasing man's power over nature, transformed the old ideas and gave birth to modern science as we know it today.

This scientific revolution blossomed in the first half of the seventeenth century, when the greatest champion of the experimental method was the Italian scholar Galileo Galilei (1564–1642). It was Galileo who first experimented with the telescope (recently invented in Holland) by looking at the stars. His observations supported the much-abused theory of Nicolas Copernicus (1472–1543) that the Sun rather than the Earth was at the centre of the universe. He also carried out careful and exact experiments on moving bodies and developed from his observations a new dynamic system of mechanics. As a result of his discoveries Galileo fell foul of the Inquisition and the papal courts, which resented his attack on long-respected authorities. This rebellion against established authorities has remained one of the most important aspects of modern science.

The leading supporter of the idea that science should be used to increase the power of man over nature was an Englishman, Francis Bacon (1561–1626). He was a contemporary of Galileo and a prominent figure at the court of King James I (1566–1625). He became Lord Chancellor

Below. Francis Bacon. He believed that science should be used for the practical benefit of man. The Royal Society was set up by his followers and was based on his ideas.

Below. Crane Court, Fleet St., London, home of the Royal Society from 1710 to 1782.

as Lord Verulam, but later fell from favour and died in disgrace in 1626. However, the next generation of English scientists continued to study Bacon's ideas in the Royal Society, founded in 1622. Among the members of this society were Robert Boyle (1627–91), Robert Hooke (1625–1703), Christopher Wren (1632–1723), and Isaac Newton (1642–1727). Under their influence the scientific revolution was established in Britain, and a start was made in fulfilling Bacon's ambitions by encouraging research in such practical matters as meteorology and map–making. Among the new ideas discussed by the Royal Society was the possibility of harnessing the power of steam.

Several lines of scientific research converged to make the exploration of steam power possible. First was the discovery, unexpected because it had never been noticed, that the atmosphere has weight. This was observed in 1644 by Evangelista Torricelli (1608–47), a pupil of Galileo, two years after his master's death. It was soon demonstrated that the air's weight varies according to its height above sea level. Second was research into the nature of a vacuum, made possible by Otto von Guericke (1602–86). He devised an air pump with which he emptied a cylinder of air and showed the effect of atmospheric pressure on a piston above the vacuum. The air pump was further developed by Boyle and Hooke, and by the Huguenot refugee Denis Papin (1647–1712), who left Paris in 1675 to take up a post under Boyle. The third significant development was Papin's invention of a method of condensing steam in a cylinder to achieve a downward stroke with a piston. This was done by boiling water (and thus making steam) to force the piston upwards, and then allowing it to cool to form a vacuum. But Papin did not think of reproducing the strokes in rapid sequence to convert his cylinder into a useful engine. Yet, with this invention in 1690, the scientific knowledge necessary for the construction of the first successful steam engine had been acquired. All that remained to be done was to convert these developments into a workable machine.

Credit for this belongs to Thomas Newcomen (1663–1729), even though he was forestalled by Captain Savery's patent which prevented him from benefiting financially from his invention. It is virtually certain that

there was a direct link between the scientific research of atmospheric pressure and steam power and the invention of the steam engine. This must be stressed as it has been argued that there was little, if any, connection between the "scientific men" of the Royal Society and the "practical men" who built the steam engine—and that Newcomen was only an ignorant mechanic who happened to stumble on a good idea. Disappointingly little is known about Newcomen, but we do know that he was a blacksmith in Dartmouth. He was literate (some letters to his wife survive), and he was a non-conformist. This tells us that he belonged to that hard core of the British puritan community who were to provide the leading figures of the Industrial Revolution. The dedication and persistence of these men largely explains why Britain led the world in the process of industrialization. Despite his humble origins, therefore, it can be said that Thomas Newcomen could have been familiar with the scientific thinking on steam power, and made good use of it.

Above. An illustration of Otto von Guericke's measurement of atmospheric pressure. By emptying a cylinder of air, a vacuum was created beneath the piston and the weight of the atmosphere was strong enough to push the piston downwards and lift the weights.

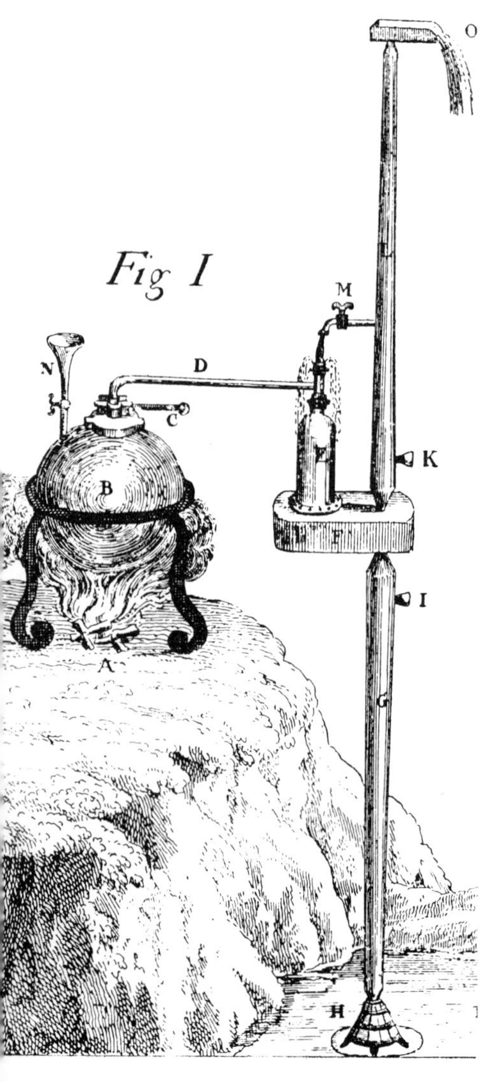

Below. An engraving of Captain Thomas Savery's "new Invention for Raiseing of Water . . .", which was patented in 1698.

At the beginning of the eighteenth century Britain had few obvious advantages over its larger continental neighbours. And in terms of population and wealth it had obvious disadvantages. France was still the dominant European power, and only a visionary could have foreseen the leading part which Britain was to take in the processes of industrialization. However, Britain did have some advantages. After the "Glorious Revolution" of 1688, it had a stable political system and a comparatively liberal social order. It was also building up a vast colonial empire and exceptional mercantile wealth. Its policy of sympathy towards the new scientific studies and tolerance of religious dissent were to bring huge commercial profits. Britain, in short, was better equipped to make rapid industrial progress than a casual comparison with, for example, France would have suggested in 1700. One of its greatest assets was the encouragement it gave to those individuals who were experimenting with steam power.

The first success was that of Edward Somerset, Marquis of Worcester (1601–67), who, in 1663, described his own engine which he claimed could raise a fountain of water forty feet by steam pressure. But his description is too confused to give a clear idea of how it worked, and it does not appear to have been copied. The first steam engine to have any commercial success was the one which Captain Thomas Savery (*c.* 1650–1715) patented in 1698. It was a "new Invention for Raiseing of Water and occasioning Motion to all Sorts of Mill Work by the Impellent Force of Fire." Savery was an adventurer and captain of mines in

Devon and Cornwall. The engine he devised (probably after reading the Marquis of Worcester's description of his invention) was intended to overcome the problem facing mining at that time—how to get rid of water. For this reason Savery referred to his engine as *The Miner's Friend,* and as most of the early steam engines were built for use in mines this is an appropriate description for all of them.

Savery's engine was, from a mechanical point of view, extremely simple. Steam from a boiler was admitted into a closed vessel. Cold water was then poured onto the outside of this vessel to cool the steam inside and to create a vacuum. This vacuum caused water to be forced up into the vessel. When more steam was admitted this water was pushed out through the delivery pipe. Once this had been done the cycle was complete. Savery's second, improved engine had two vessels which discharged alternately to give a continuous flow of water. However, the engine was not very effective because water can only be forced up to a height of twenty-eight feet, and because it relied on steam pressure to push the water out. This reliance upon high steam pressure (at a time when boilers and water containers were still extremely primitive) and the loss of energy caused by bringing the steam into contact with cold water, made the Savery engine expensive and unreliable. It could only produce short lifts, and relied mainly on the condensation action.

The first successful mine-pumping engine was developed at the beginning of the eighteenth century. It fell within the terms of Savery's patent so that he and his executors managed to get financial control over it. This was Thomas Newcomen's engine which used Denis Papin's cylinder and piston, and repeatedly injected and condensed steam to create a partial vacuum. The piston was coupled to a swinging beam which drove the pump rods. It also had an automatic control for the valves, worked by a rod on the swinging beam. In the design of this engine Newcomen displayed the genius of a true innovator.

The first full-scale Newcomen engine was built near Dudley Castle in Staffordshire in 1712, to work in a coal mine. Both Savery and Newcomen began by considering the need for pumping engines in the mines of Cornwall and Devon. But by far the most significant use of the New-

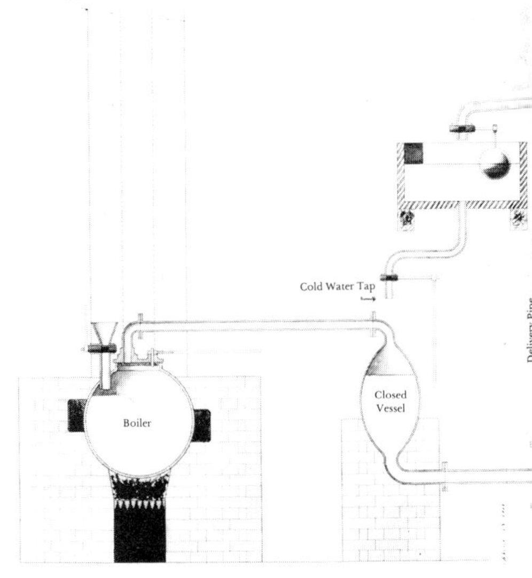

Below. A diagram of Savery's first engine which he called *The Miner's Friend.* It was not very efficient, mainly because cold water came into direct contact with the steam, and the boilers were too weak to provide the necessary steam pressure.

comen engine was in the coal mines of the Midlands and northern counties. The engine used a lot of coal, which meant that in areas where coal was difficult to obtain, the cost of operating such a machine was too high. In the coalfields, however, the cost of coal was low and the engine only used small coal which would otherwise have gone to waste. So in these areas the Newcomen engine was cheap to run and proved to be the coal-miner's friend; hundreds were built during the eighteenth century. Many were so robust and reliable that they operated with the minimum of maintenance for many decades.

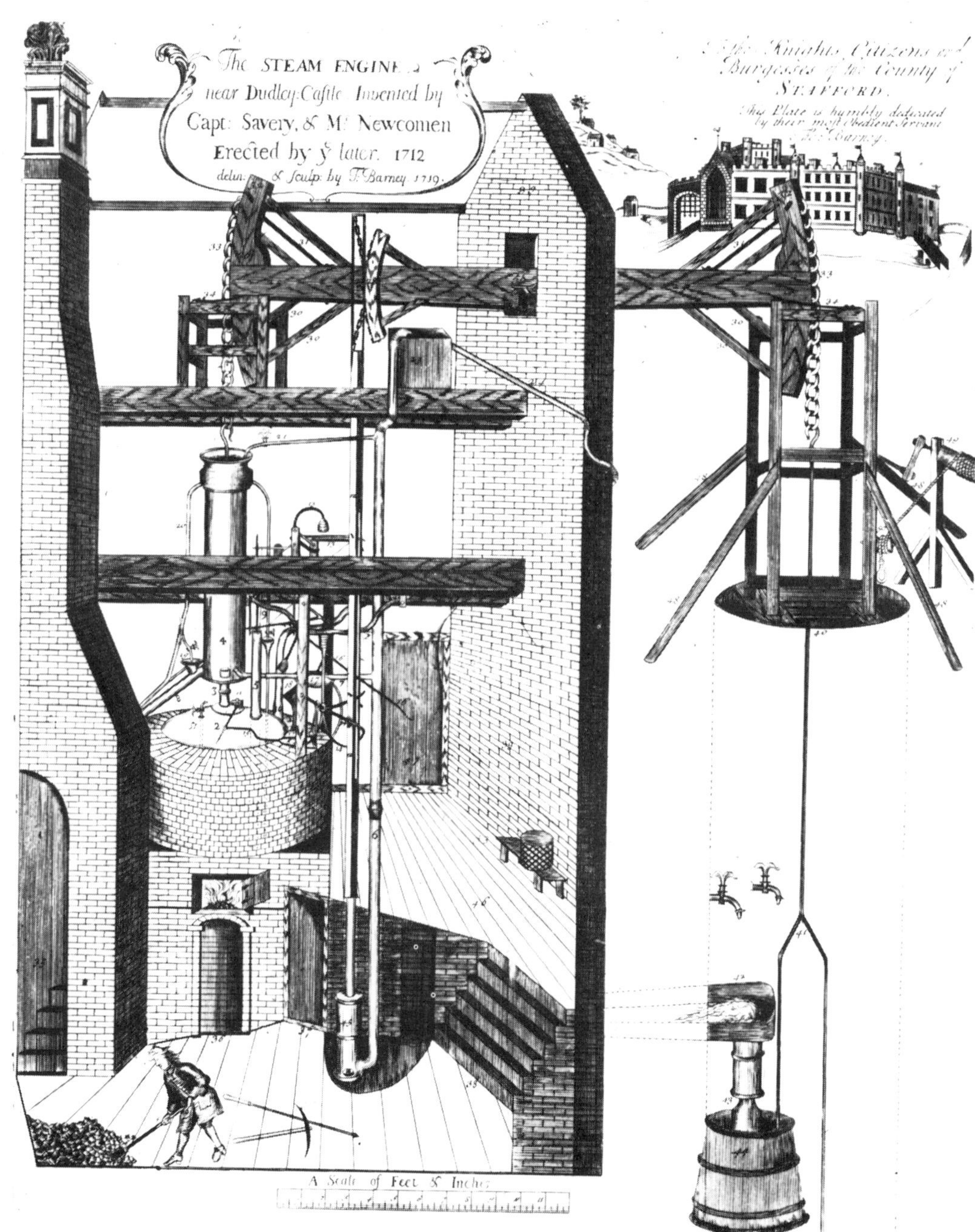

The Newcomen engine answered a specific social need—the rapidly increasing demand for coal. As it burnt coal it, too, added to the demand for greater coal production as well as stimulating the industries which made the large iron cylinders and other parts of the steam engine. The cylinders of the earliest engines had been made of brass—which was easier to work than iron—but iron was stronger and cheaper and was gradually used for most of the metalwork in the engines. The Newcomen type of steam engine could therefore make use of existing resources of materials and capital, and could be built by the skilled blacksmiths and mill-wrights of the period. Its success led to many refinements in construction and minor improvements by Henry Beighton (*d.* 1743), John Smeaton (1724–94) and other engineers. But there was no radical change in Newcomen's basic design until James Watt (1736–1819) patented his separate condenser in 1769.

The early steam engines were called "fire engines" because people were impressed by the use of fire to produce power. The action of the steam or "rarefied air" remained mysterious to most people throughout the eighteenth century. Apart from Savery's claim to use steam under pressure (which was abandoned in favour of relying almost entirely on condensation) virtually all the early engines were atmospheric—that is, condensing steam created a partial vacuum which allowed atmospheric pressure to provide the power stroke. It was only in the engines of James Watt that the change from atmospheric to expansive steam action was made, and that was just a beginning because Watt continued to rely on the condensation of steam as his main source of power.

Watt was interested by the ability of steam to behave as "an elastic body." It could expand to occupy many times the volume of the water from which it came, and could then contract to leave a vacuum. Or it could be drawn into a vacuum by the opening of appropriate valves. Thinking about this elasticity of steam, he had an inspiration one Sunday morning in May, 1765, while strolling on Glasgow Green. He had been employed as instrument maker by Glasgow University and had been trying to make a model Newcomen engine work more efficiently. His work on the properties of steam gave him the idea of separating the

Above. A cut-away model of the New-comen mine-pumping engine which was built at Fresnes in France in about 1735.

15

Above. An artist's impression of James Watt in Glasgow studying Newcomen's engine to find a way of making it more efficient.

condenser from the cylinder in which the power stroke was made. By keeping the condenser permanently cool and the cylinder permanently hot, he realized he could avoid the waste of energy involved in heating and cooling the cylinder on every stroke, as in the Newcomen engine. This was the invention for which, four years later, he took out the separate condenser patent, and which became the basis of his successful partnership with Matthew Boulton (1728–1809) between 1774 and 1800. In its first form, in 1769, Watt's engine was still an atmospheric machine, but he quickly discovered that to keep the cylinder hot he had to protect it with a steam jacket and thick lagging. He therefore covered the top of the cylinder, which had been left open in the Newcomen engine. Now instead of the atmosphere acting upon the top of the piston, Watt used steam at about atmospheric pressure. In 1784 he invented the "double-acting" engine. Having done away with the intervention of atmospheric pressure it became possible to introduce steam and to condense it both below and above the piston alternately, controlling the process by an automatic sequence of valve changes.

From this brief account it is clear that the Watt engine was much more advanced than the Newcomen type, and this caused its difficulties. Greater efficiency was acquired, but the initial installation was much more expensive. It also needed more expertise in manufacture. Newcomen engines had been built on site (as far as possible with local labour and materials) and this practice was continued by Boulton and Watt. But the finely machined valves and other moving parts were made in their own workshops at Soho, in Birmingham. This use of precision engineering techniques for the manufacture of large machines gave birth to the modern engineering industry. Up till now such skills had been confined to clock-making and the manufacture of scientific instruments. By making them standard practice in engine making, and combining them with the traditional skills of the millwright, the Soho firm revolutionized industrial organization.

The problems involved in this innovation were considerable. The main difficulty was cost. The Watt engine

Below. James Watt's workshop in Soho, Birmingham. All the valves and other finely machined parts of the Boulton and Watt steam engines were made in the Soho factory.

was an expensive substitute for a Newcomen machine, and there was little incentive to use it in the coal mining industry, where fuel was cheap. Boulton and Watt realized, however, that there was still a need for a good pumping engine in the tin and copper mines of Cornwall and Devon, and it was here that they sold their engines. Metal mining in the south-west underwent a great revival with the help of the new engines. But Boulton and Watt were not popular for, in an effort to recover their heavy investment costs, they thought up a form of payment based upon the amount of fuel saved by using their engine instead of a Newcomen-type engine. The mine-owners objected to the supervision and payments of these sums. When they had the opportunity to adopt the high-pressure steam engine

Below. The massive beam of the Boulton and Watt pumping engine which was installed in 1777 for work on the Birmingham canal.

invented by the Cornishman, Richard Trevithick (1771–1833), the Cornish mines captains did so with relief. But they kept a strong sense of economy—engines were closely observed for their performance and efficiency long after Boulton and Watt had retired. However, it is clear that James Watt was as much the "miner's friend" as Savery or Newcomen had been before him.

By 1800 the urgent need for better pumping machines in both the coal and metal mines of Britain had therefore been met. Furthermore, in overcoming the technical problems inventors and engineers had discovered a machine with enormous potential. But its potential was not fully recognized until the closing decades of the century for two reasons. Firstly, the large and inflexible Newcomen engine was ill-adapted for any other use. Secondly, there were few industrial applications for steam power until the coming of the factory system. It was this new form of industrial organization which created the pressing need for an invention which would free industry from dependence upon water power. At precisely the moment when the steam engine had met the needs for which it was originally designed, many new needs appeared.

Left. James Watt, one of the greatest pioneers of steam power.

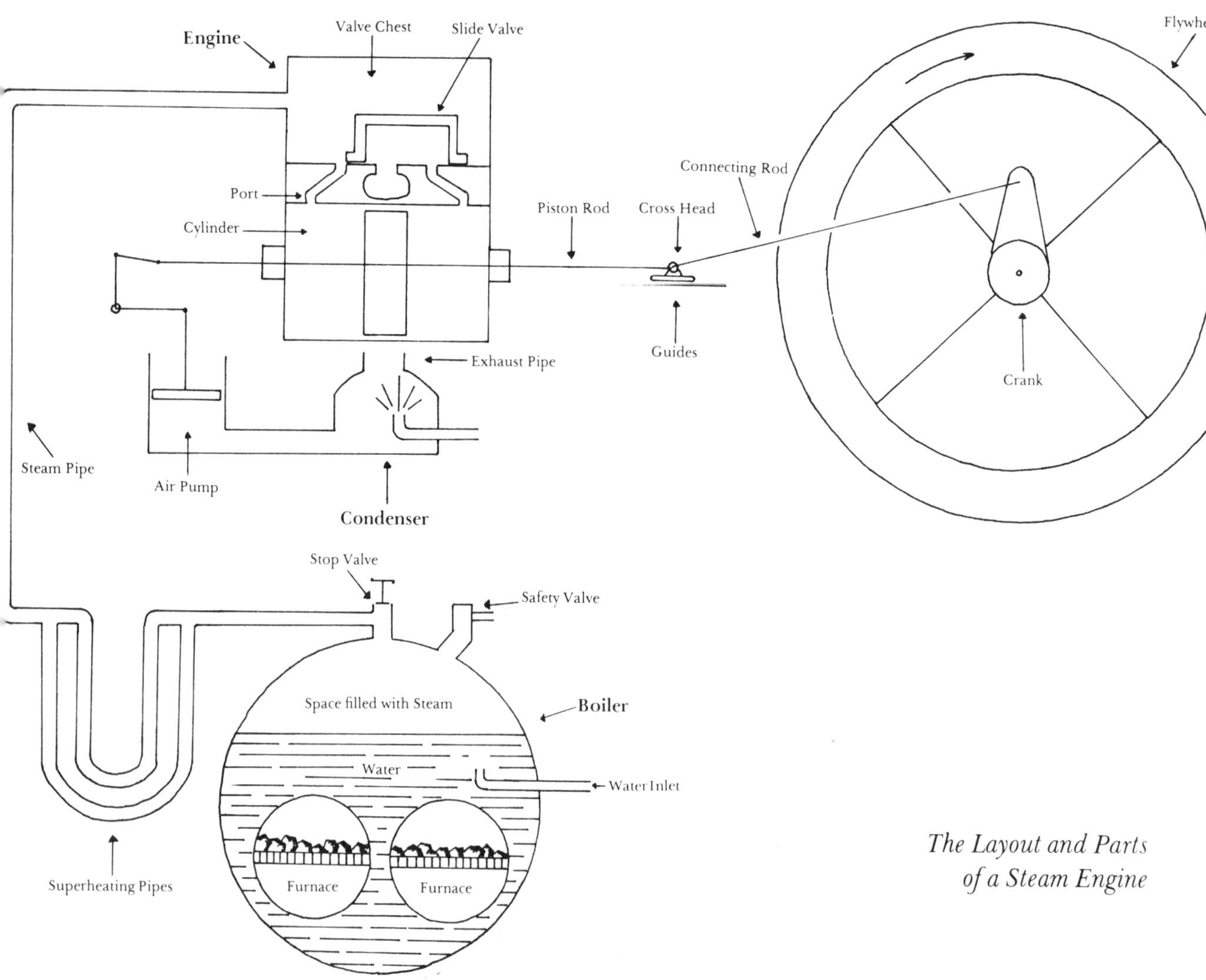

*The Layout and Parts
of a Steam Engine*

A complete steam power system consists of three parts—a boiler to generate the steam under pressure, an engine to produce power from the steam, and a condenser to convert the used steam back to water for return to the boiler and to make a vacuum to help the engine.

The Boiler. Steam is produced by heating the water in the partly filled *boiler*. In this example, the two *furnaces* contain heating fires and the steam fills the space above the water. As more and more steam fills this space there is a rise in pressure and temperature. The steam gets to the engine through the *stop valve* and the *steam pipe*. (A *safety valve* on the boiler releases steam when it exceeds the allowed pressure.) It is the pressure rise which enables the heat of the steam to be used as energy in the engine. The heat causes the steam to expand, and give up the energy which then powers the crankshaft and the *flywheel*.

The Engine. The steam from the boiler passes through the steam pipe to the *valve chest*. The valve chest contains a *slide valve*, which is a cast iron box that slides steamtight upon a flat face. It has openings (called *ports*) which let steam onto the *piston*, as well as releasing it after it has done its work. The piston slides freely, without any leakage, within the circular bore of the *cylinder* and is connected by the piston rod to the *crosshead*. From the crosshead the thrust is transferred to the *connecting rod* which in turn rotates the *crank* and the flywheel. The crosshead is kept parallel by guides which also restrict the sideways motion of the connecting rod.

The Condenser. Once the steam has worked the piston it has to be let out of the cylinder. It can either pass out into the air through the *exhaust pipe,* or (as in the locomotive) into a chimney to increase the draught. These are simple but wasteful methods. A better solution is to let the exhaust steam into a *condenser,* which can recover the water from the steam. By producing a vacuum it also makes the engine more efficient. Since it has to remove a great deal of heat, condensing requires a lot of cold water for cooling the exhaust steam. This is done in cooling towers which are the large chimney-like structures we see discharging vapour at the power stations. Condensing also needs a small *air pump* to remove the uncondensable gases that would otherwise spoil the vacuum.

Steam engines were mainly built of cast iron, though steel was used for the highly stressed parts. The parts which rubbed against each other were fitted with "brasses" (as they were called) but which were in fact bronze bearings that could be adjusted when they became worn. A lot of fuel was saved if the steam was superheated: that is, its temperature was raised by passing it, on its way to the engine, through pipes heated by the hot gases from the boiler. This meant that the steam did not begin to condense until it had lost the heat which had been added in the *superheating pipes*.

Newcomen's Engine: The Working Cycle

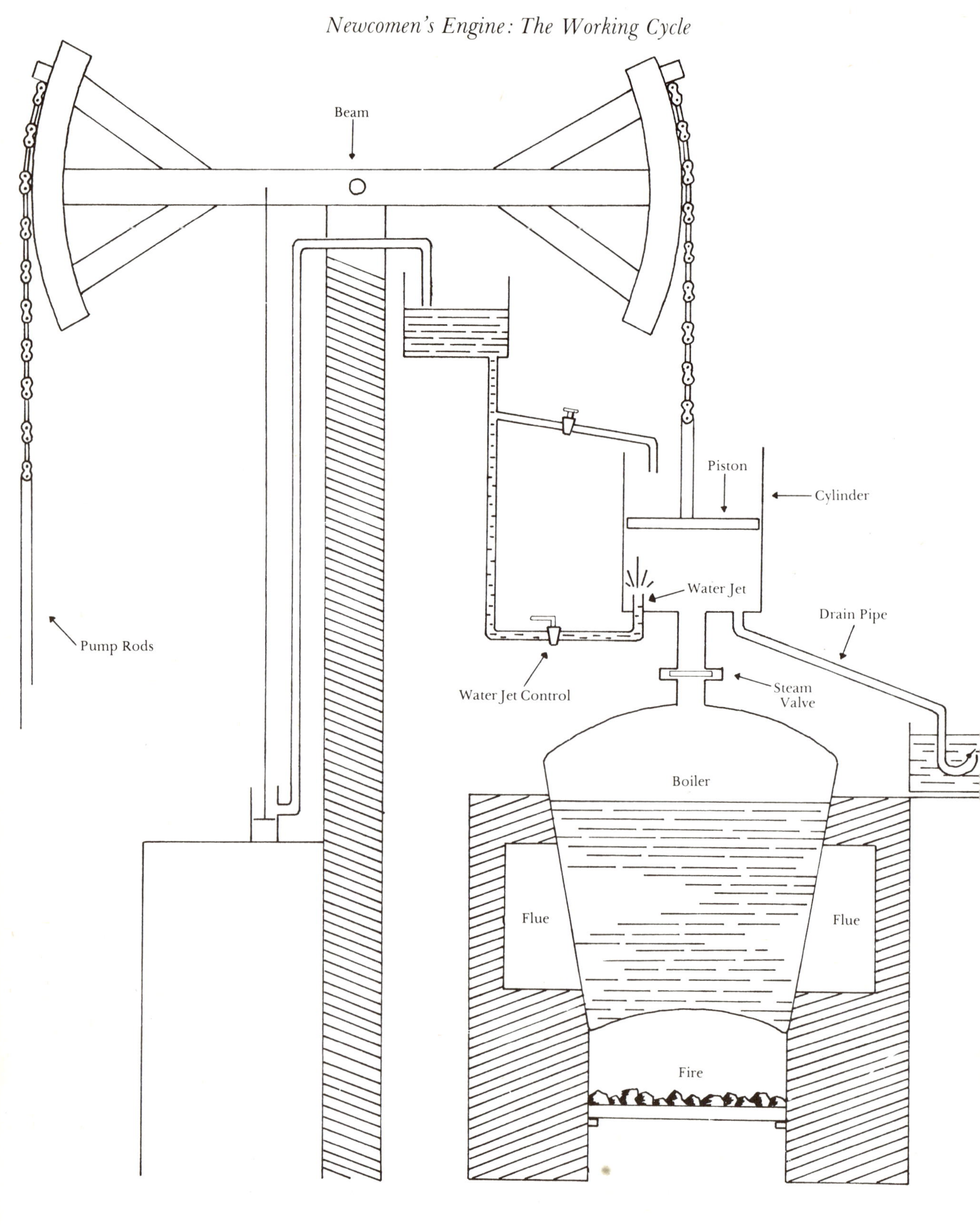

The water in the *boiler* was heated by the fire lit below it, and by the hot gases passing through the *flues* on the way to the chimney. Steam at atmospheric pressure was admitted below the *piston* by the *valve* when the piston was at the bottom of the stroke. This equalized the pressure above and below the piston, which allowed the weight of the *pump rods* in the pit to pull the piston to the top of the *cylinder*. The steam was then shut off and a jet of cold water was sprayed into the cylinder steam which condensed to produce a vacuum beneath the piston. The pump rods were then raised as the vacuum drew the piston downwards in the cylinder.

Newcomen's engine was simple and easy to build, but it wasted a lot of steam because the cylinder was cooled by the water injected into it. Therefore much of the steam was condensed as it entered from the boiler. This was the great failing of the first working steam engines. The design also meant that it was necessary to keep water on the top of the piston to prevent air leaking in to spoil the vacuum. The condensed steam and the cold water injected drained away from the cylinder by the *drain pipe* as soon as the fresh steam was admitted. This could happen because the vacuum was spoilt once fresh steam came into the cylinder.

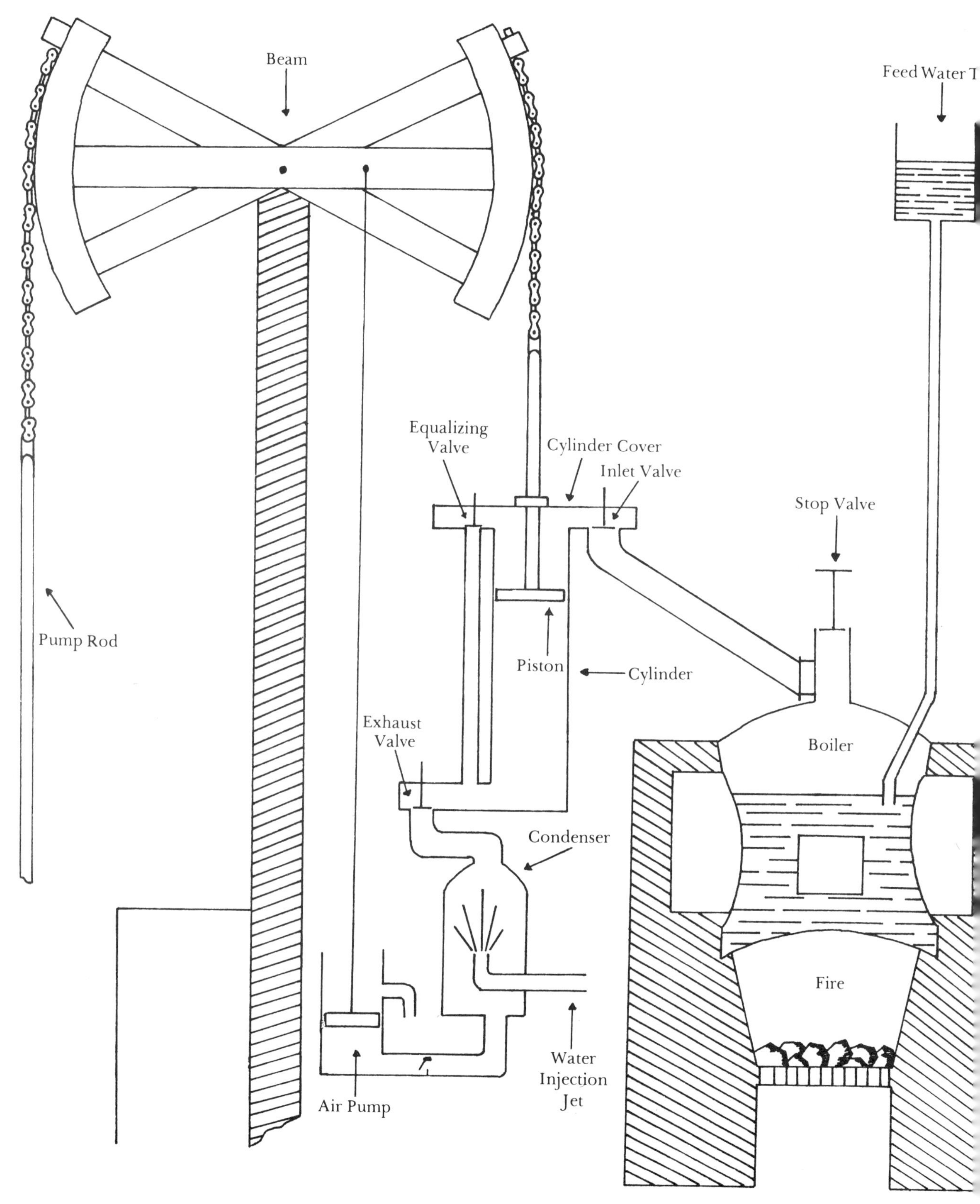

James Watt's Engine with Separate Condenser
Beam
Feed Water T
Equalizing Valve
Cylinder Cover
Inlet Valve
Stop Valve
Pump Rod
Piston
Cylinder
Exhaust Valve
Boiler
Condenser
Air Pump
Water Injection Jet
Fire

In Watt's engine, when the steam from the *boiler* entered the top of the *cylinder* (which was closed by the *cylinder cover*) through the *inlet valve,* the *exhaust valve* was open to the *condenser*. The combined effort above and below the *piston* raised the *pump rods* as the piston moved to the bottom of the cylinder. The inlet valve and exhaust valve were then closed, and the *equalizing valve* was opened. The steam above the piston then passed through the equalizing valve and the side pipe to the underside of the piston. This equalized the pressure on both sides of the piston which meant the weight of the pump rods now pulled the piston to the top again ready for another cycle to begin. In this way the cold *condenser* was completely isolated from the cylinder, except when the vacuum was needed to help the piston raise the pump rods. Condensing, therefore, takes place in a separate chamber cooled by a *water injection jet*. But since the condenser is now a vacuum, the condensed steam and injection water cannot drain away as in the Newcomen engine. So the water, and the small amount of air that leaks in, are now removed by a separate, small *air pump*. The cycle is completed when the water, which was evaporated into steam, is pumped to the *feed water tank*. From there it returns to the boiler by gravity.

3:   *Early Industrial Uses*

The factory system was already going strong before James Watt invented the rotative action which allowed steam power to be harnessed to spinning machines, looms, rolling mills, and countless other industrial machines. Factories had appeared first in the textile industries, where cheaper goods were produced by putting all the machines and workers in one set of buildings under one owner. Because they relied mainly on water power, these factories had been built in places where there was a good source of water. And so they were frequently situated away from the major centres of population—like Richard Arkwright's mills at Cromford near the River Derwent in Derbyshire, and the mill community set up by Robert Owen (1771–1858) at New Lanark on the Falls of Clyde. The steam engine released the developing mills and forges of the Industrial Revolution from this dependence on water power, and encouraged their construction in ports, towns and coal fields. This was the greatest relocation of industry in British history, and the steam engine was responsible for it.

Until the 1780s, the steam engine had been basically a pumping machine for removing water from deep mines. Occasionally engines were adapted for other pumping duties, but their usefulness was severely restricted by their simple reciprocating (or "back-and-forwards") action. To convert this action into rotative movement, the steam engine was first used as a pump to fill a reservoir with water which then drove a conventional water wheel. There were numerous installations of this sort, such as the Newcomen

Opposite. The water-powered cotton mills of Robert Owen's New Lanark mill community in the early nineteenth century.

Above. Boulton and Watt's engine with rotary action. The "sun and planet" gearing is seen on the right, and the interconnecting rods, which keep the piston vertical ("parallel motion"), are visible above the cylinder.

engines built by William Champion in his brass works at Warmley, near Bristol. By the 1780s, there was an intensive search for a satisfactory method of producing rotative action from steam engines, as industrialists began to realize the possible advantages of steam power. Some experimental models used cranks to convert the reciprocating action into rotary action, but as these were used with rather poor engines their performance was disappointing.

This then was the situation in 1781 when Matthew Boulton urged his partner to develop rotative action, and Watt complained that "the devil of rotations is afoot." Anxious to avoid using the crank, on which rivals had already sought a patent, Watt developed his "sun and planet" gearing. A gear wheel was fixed to the end of the connecting rod (the "planet"), which moved round a similar wheel (the "sun") on the driving shaft. As the connecting rod moved up and down the "planet" moved round the "sun" and gave the driving shaft a rotary motion. Watt patented this arrangement in 1781, and with a heavy fly-wheel on the driving shaft the engine produced a smooth rotative action. In the next few years he solved the remaining problems of this new type of engine. In 1783 he made use of "double action," (producing a power-stroke in each direction) to further increase the smoothness of the rotary action. In 1784 his "parallel motion" patent of 1784 enabled the piston rod to be kept vertical throughout its movement by linking it to the swinging beam by an elegant parallelogram of inter-connecting rods. In 1787 his adaptation of the "centrifugal governor," by which the spinning action of two weights adjusted the steam admitted to the engine, introduced a reliable control on the rotative action.

By the end of the 1780s, therefore, Boulton and Watt were manufacturing a completely satisfactory form of rotary action steam engine. These were quickly bought by those industrialists who had an eye to innovation and rapid development. John Wilkinson (1728–1808), the Midlands iron-master who had bored the cylinders for the early Boulton and Watt engines to a unique standard of ac-curacy, installed one of the first rotative action engines. They were soon being widely used in the iron industry both as blowing engines for blast furnaces and for rolling mills in forges. A set of rotative action engines was installed in the specially built Albion Mill alongside the River Thames at Blackfriars in 1784 as a source of power for corn-grinding. The steam engine was now taking over what for centuries had been done by wind and water power.

More significant for the future was the introduction of the rotative action steam engine into the cotton textile in-dustry. Although Richard Arkwright (1732–92) had built most of his mills in the Derwent valley of Derbyshire where water was plentiful, he had experimented with simple rotary action engines made by Wrigley in Manchester in 1783. But he was not immediately convinced that they could be used to spin good thread on his spinning frames. Other cotton manufacturers were more willing to turn to steam power. Boulton and Watt engines were not the only ones used. Newcomen machines were also being adapted

Above. A close-up view of the two weights (the "centrifugal governor") which adjust the amount of steam ad-mitted to the engine. The mechanism which drove the valves can also be seen.

Below. Richard Arkwright, the first mill owner to experiment with rotary action steam engines in factories.

for rotative action. These "common" engines (as they were known in the 1790s) cost only £200 compared to about £850 for a Soho machine so cotton manufacturers had a good reason for trying to make do with the less efficient engine. But the greater efficiency of Watt's engines gave better results, and as the "common" engines were also notoriously dirty at a time when town dwellers wcrc beginning to worry about air polution, the comparative cleanliness of the Boulton and Watt machine gave it a great advantage.

Steam power did not take over the British cotton textile industry immediately. Many large water-powered factories of the pre-steam era remained prosperous without turning to steam power until well into the nineteenth century. A few such mills even continued to use water power, at least for some purposes, up to the present century. But all the *new* development in this rapidly expanding industry used steam power after 1790. Steam power meant that factories could be sited anywhere. Many therefore moved to the towns associated with the old hosiery industry, like Nottingham, or to the mill towns of south Lancashire, with their easy access to Liverpool and the Lancashire coal field.

Below. An engraving by the famous English artist William Hogarth, illustrating the hardship of working in the cotton industry in the eighteenth century.

Left. An atmospheric steam engine, of the type which enabled British coal production to increase dramatically, installed at the Moira colliery in Leicestershire.

Fed by coal, steam and imported cotton, the industry flourished in these areas and became the outstanding growth achievement of the first half of the nineteenth century.

Apart from the cotton industry, the steam engines made a considerable contribution to the development of other industries in the first half of the nineteenth century. The use of steam power in the iron and steel industry has already been mentioned, and this grew steadily in the nineteenth century. It speeded up the transfer of the industry from the backwoods of the Weald, Dean and Shropshire, to the coalfields of the Midlands and south Yorkshire, where there was both an urban market and an urban pool of labour. The usefulness of the steam engine to the mining industry should also be remembered. It had always been the "miner's friend" in that it had freed the mines of water by providing an efficient pumping service. But with the coming of rotative action, the steam engine could meet other needs of the mining industry, especially for winding engines and for ventilating machinery. The British coal industry in particular benefited from many types of steam power and its production figures rose up and up to an all-time peak in 1913.

In every industry, where wheels had to be turned regularly and reliably, the rotative action steam engine was put to use. The new engines were installed in the paper making, newspaper printing, pottery manufacture, brewing, and tanning industries. The "devil of rotations" was indeed afoot, and it seemed to many sensitive observers that the steam engine had brought in the era of "the dark Satanic mills."

The steam engine was a powerful agent in promoting the spread of the factory system. But it brought little benefit to the ordinary factory worker. Most people lived hard lives, ruled by the factory clock, the tireless rhythm of the steam engine and the fear of unemployment. By modern standards, life was hard in the Industrial Revolution, with no comforts and few, even, of the necessities of life. But there was a great achievement in these harsh conditions. Steam technology, and the whole complex of industrial mechanization of which it was a part, made possible an unparalleled increase in productivity. This enabled a rapidly rising population to be kept at a standard which, however poor it may seem to us, held the promise of eventual improvement. In this achievement the story of man and the

Below. The "dark satanic mills" of the Industrial Revolution, which some people thought were the direct result of the introduction of the steam engine.

steam engine reaches its highest peak.

Society now looked to steam power to meet its increasing needs. The remarkable success of the Industrial Revolution in terms of increased productivity and cheap commodities for the worker may be largely attributed to the steam engine. However, the growth of possible uses for steam technology forced engineers and inventors to improve on Watt's inventions. When the Watt patents lapsed in 1800, both James Watt and Matthew Boulton retired from active participation in their firm, which was then taken over by their sons. James Watt's retirement resulted in a burst of activity in those areas where he had most resisted innovation—high pressure steam technology and steam transport. Of these, high pressure steam was the more important as it was itself vital to steam transport. It also added a new dimension to the industrial uses of steam power.

Richard Trevithick was the master-mind of high pressure steam technology in Britain. But his achievements were closely paralleled by those of Oliver Evans (1755–1819) in the United States. Watt had already experimented with the expansive energy of steam, having provided for it in his 1784 patent. But as his boilers could not make steam safely at more than four or five p.s.i. (pounds per square inch) the possibilities for this type of steam engine were severely limited. So he continued to rely heavily upon the condensation action until 1800.

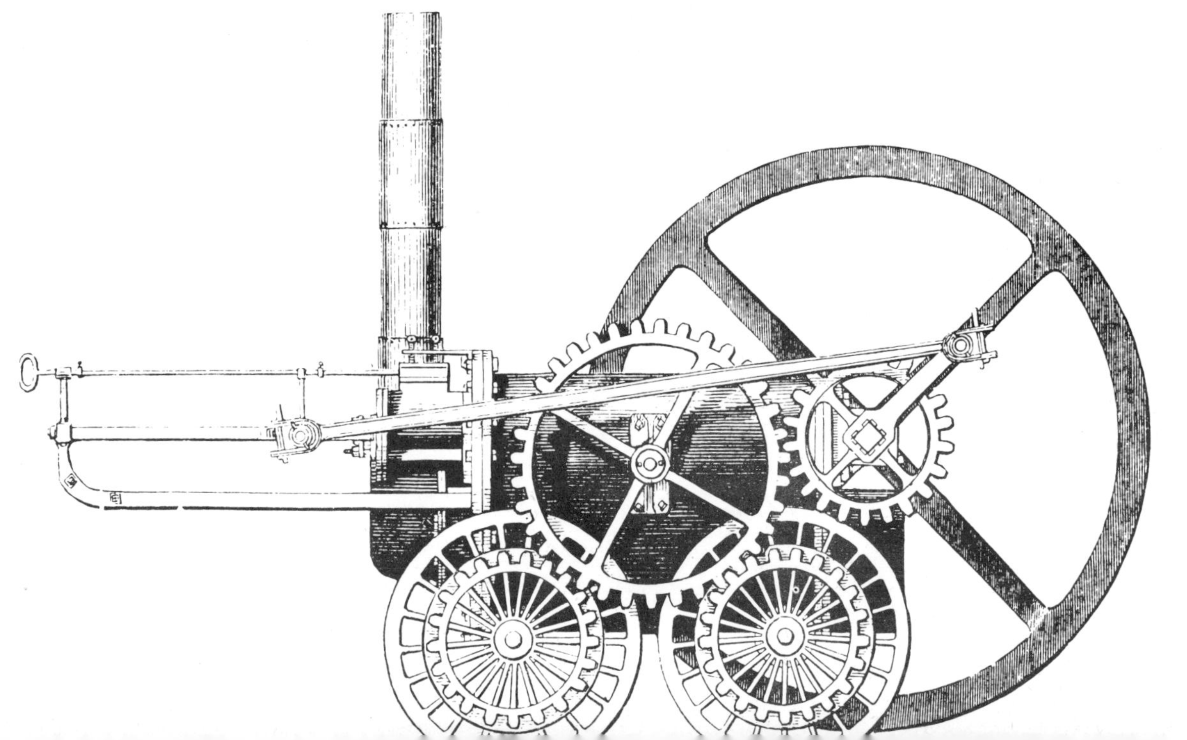

Below. Richard Trevithick's high pressure steam locomotive which worked on the Penydarren Tramway in south Wales in 1804.

Trevithick did not share Watt's inhibition, and set out to make a boiler strong enough to support a pressure of fifty p.s.i. Despite some explosions and near tragedies he succeeded, and was able to drive a remarkably compact engine by using only the expansive power of the steam—thus dispensing completely with the condensing action. He was quick to realize the possibilities for transport of such an engine, and at Christmas 1801 fitted it to a road carriage near his home at Camborne, Cornwall. So began the era of the steam locomotive.

High pressure steam brought greater efficiency, greater compactness of design and greater mobility to the steam engine. As they were no longer dependent upon large engine houses, many steam engines became almost self-contained. There was a considerable variety in the size and

Below. A "grasshopper" beam engine, a compact design useful for smaller workshops and pumping stations.

shape of steam engines in the first half of the nineteenth century. Compact designs (many of them without the traditional swinging beam) such as "grasshopper engines," (with the beam pivoted at one end instead of the centre), "table engines" (so called because the vertical cylinder was supported on four columns resembling a table), and oscillating engines (in which the piston rod from a pivoted cylinder drove directly onto the crank) were introduced. By the 1820s, high pressure engines with horizontal cylinders were beginning to find their way into industrial use. But high pressure could also be applied to the traditional "house-built" beam engines, and in the deep mines of Cornwall in particular this was done with great success. By combining high pressure steam and expansion with condensing action, an engine of outstanding power and efficiency was produced. It was known as the "Cornish" type engine even though they were widely used outside Cornwall.

Expansive steam action had only come about indirectly, as inventors had been limited by the constructional methods of the eighteenth century. But a hundred years of experience with the atmospheric and condensing engine brought increasing confidence and skill in boiler making. So it became possible eventually to take full advantage of the elastic properties of steam by using its expansive action in high pressure engines. At once, exciting new possibilities opened up for the high pressure steam engine in both its stationary and mobile forms, and inventors began to exploit these possibilities.

Left. A "table" engine from the first half of the nineteenth century. This was one of the first types of steam engine to get rid of the swinging beam.

Watt's Engine with Rotative Motion

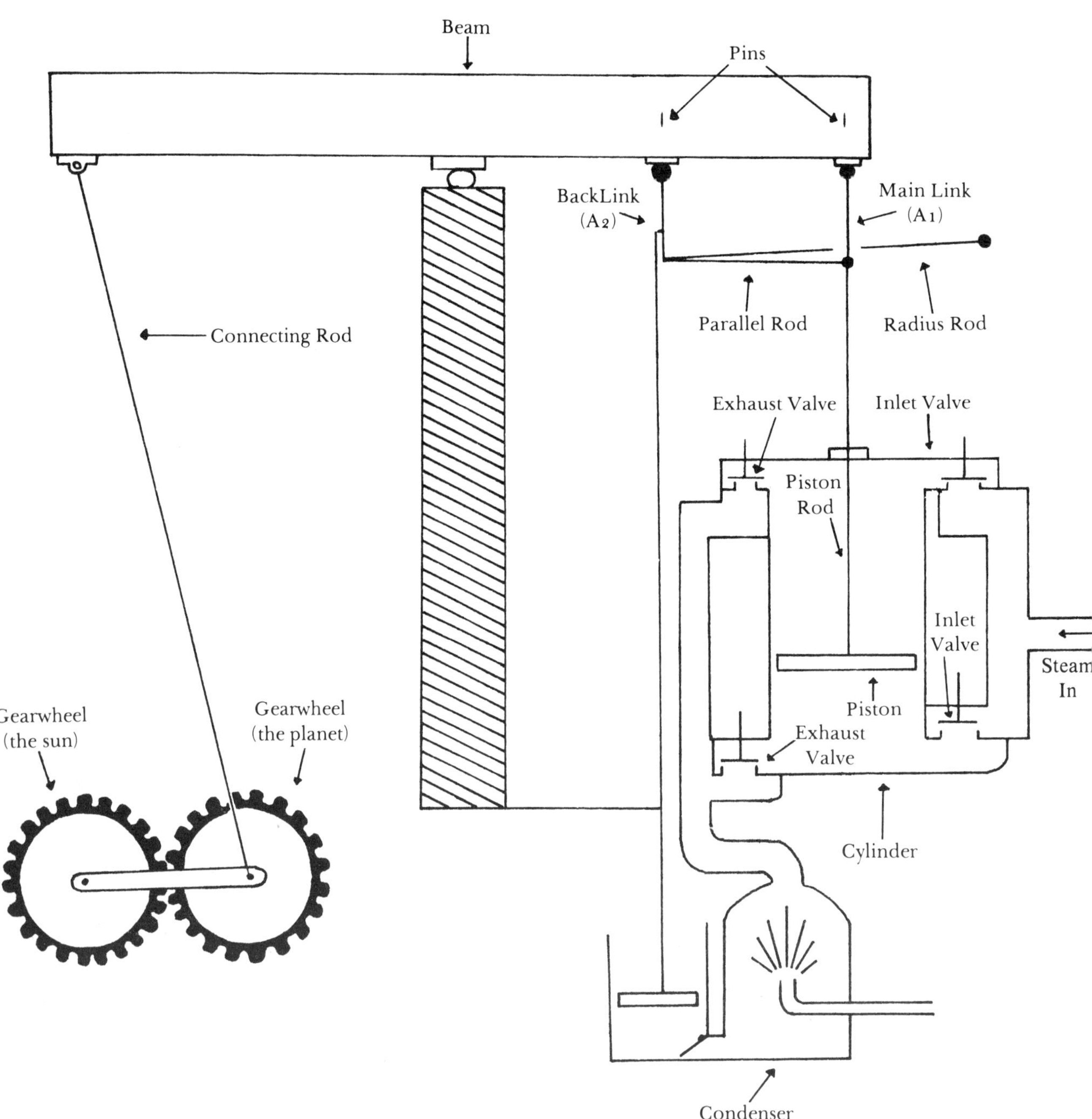

Watt's "sun and planet" method of obtaining rotary action makes use of two gearwheels. One (the *planet*) is rigidly fixed to the *connecting rod*. The other (the *sun*) is fixed to the *crankshaft*. The two gearwheels are meshed together. As the "planet" could not rotate the crankshaft by itself, its teeth were meshed with those of the "sun" to turn the crankshaft. Thus the "planet" moves around the "sun." This invention made the crankshaft rotate at twice the speed of a conventional crank.

In the double action engine the *piston rod* both pulls and pushes the beam. As the head of the rod must move in a straight line (while every point in the beam describes an arc of a circle), the *beam* and the rod can not be rigidly connected. To solve this problem Watt invented an arrangement of rods called "the parallel motion." He connected a main link (A_1) from the top of the piston rod to the beam. As the piston rod moved up and down this link moved slightly away from the vertical. The back link (A_2), of the same length as A_1, was placed further along the beam. *Parallel rods* (the same length as the pins in the beam) connected the bottom of A_2 to A_1 to ensure both links moved in the same alignment at all times. Finally, to guide the top of the piston rod in a straight line, *radius rods* were fitted. These were attached to pins mounted on the engine frame and fixed to the bottom of A_2. They restrain the movement of A_2, making sure that the top of the piston was guided in a straight line.

The double action engine had to admit and exhaust the steam at each end of the *cylinder* alternately. This required separate *inlet* and *exhaust valves* so that the steam could be admitted and exhausted at each end of the cylinder, and not transferred from the top to the bottom as in Watt's earlier pumping engine.

4
THE WALLIS TRACTION ENGINE
HO 5587

4: *The Steam Engine in Agriculture*

The modernization of agriculture is a central part of the process of industrialization. A growing industrial community needs a steady supply of food, and the transformation of the farming industry in Britain was so dramatic that it can fairly be described as the "agricultural revolution." The first steps in this transformation were organizational rather than technical. First it was necessary to change the pattern of British farming by setting up "enclosed" farms, which became the experimental units for new crops and techniques. Enclosure was carried out in a piecemeal way by local landowners petitioning for Acts of Parliament to redistribute property. This lengthy process was started early in the eighteenth century and was not completed until the middle of the nineteenth century.

On the enclosed farms of the more adventurous landlords many new ideas were tried out. Some introduced entirely new crops such as turnips, or tried new systems of crop rotation with manuring and other methods of keeping the soil healthy. Other innovations were made in stock breeding. By separating the herds and flocks from those on neighbouring farms, it became possible to control mating and to check on the improvements made by selective breeding. Further changes brought machinery into agriculutre, and it was here that the steam engine made a great contribution.

In the eighteenth century, when steam engines were still large machines in purpose-built engine houses and used mostly for pumping water out of mines, it was difficult to see how they could be used in agriculture. The early im-

Opposite. A Wallis steam traction engine being used to power a saw.

provements in agricultural mechanization, therefore, were in the design of ploughs, ploughshares, and the seed drill—first popularized by Jethro Tull (1674–1741). Agricultural societies sprang up all over the country to help farmers to experiment with such techniques but it was not until the "high farming" era of the middle decades of the nineteenth century, when farming had become efficiently organized and prosperous, that it was possible to think of the practical implications of steam technology for agriculture.

The steam engine was first used in agriculture for land drainage. This meant it had to pump water—a task which steam engines had been originally designed to perform. The main difference was that land drainage involved lifting a large volume of water only a few feet. Traditionally, this task had been performed by windmills, (in areas like the Fenland around the Wash or the Dutch polders) which were linked to "scoop wheels" which raised the water. The greater reliability of steam engines made them more and more attractive for this purpose and many of the larger pumping stations changed to steam power in the first half of the nineteenth century. The Dutch engineers who drained the Haarlemsmeer polder near Amsterdam in the 1840s used three massive steam engines. One of these, the Cruquius engine, has been preserved even though it was taken out of commission in the 1930s. This engine was made by Harveys of Hayle and consists of annular cylinders (the high pressure cylinder is inside the low pressure cylinder) driving eight beams from a single crosshead. These are connected to eight simple lift pumps, raising the water a few feet. A much more conventional land drainage engine has been preserved in Britain at Stretham, near Ely. This was built by the Butterley Company of Derbyshire and operated from 1831 to 1941. It drove a scoop wheel which was later enlarged to make up for the sinking of the drained levels.

Trevithick made an engine to drive barn machinery in 1812 and planned to make one which was self-propelling. But in the unsettled agricultural conditions of the early nineteenth century little progress was made in this field. Other agricultural applications of steam power were introduced gradually in the mid-nineteenth century. Steam

Below. A nineteenth-century farmyard. Steam technology was able to take over many jobs on the farm previously done by hand.

ploughing presented problems because even though steam traction engines were mobile, they were too heavy to cross open fields without damaging the sub-soil and getting stuck in bad weather. So the horse-drawn plough was not seriously challenged until a way was found of hauling the plough across the field on a cable. This was done either with a ploughing engine at both ends of the field, or with a winch at one end returning the cable to the engine at the other. A lot of money, time, skill and patience was needed, but once the machinery had been set up it worked with great speed and efficiency. To get the best results from steam ploughing the fields had to be large and regularly

Below. The unusual Cruquius engine, used for land drainage in Holland, which drove eight beams from a single crosshead.

Above. Steam ploughing in Victorian England. The plough is being hauled across a field by means of a cable linked to a winch driven by the steam engine.

shaped, as the engine and winching apparatus took up some "dead" ground at the end of the field. This "dead" ground increased as the size of the field got smaller. But for all its disadvantages, steam ploughing was widely adopted on the large farms of south-east England. And despite the hardships of the agricultural depression after 1870, which mostly affected the large arable farms with a heavy financial investment in machinery, steam ploughing continued until the 1920s when it was superseded by the internal combustion tractor.

The steam plough was a type of traction engine. Like a locomotive it had its own boiler and firebox, and it could also work as a power source for other machines. In this form the steam engine was linked to a variety of machines on the large farms of mid-Victorian England. Chaff-cutters, root-pulpers, threshing machines, elevators and many other uses for steam engines were found on the

Above. Steam engines greatly contributed to the "agricultural revolution" of the nineteenth century. Here, a portable steam engine is providing power for the threshing and stacking machine.

mechanized farm. They were used, too, in saw-mills and for other purposes related to agriculture. And the mobile traction engine could even be used for heavy-duty haulage work.

It is clear therefore that the agricultural industry in Britain benefited from steam power. In other parts of Europe, however, farming remained a poor industry and there was no investment in steam power. Yet for the prairie economy of North America, which came to supply more and more grain for the British market, mechanization was a major reason for its competitiveness and cheapness. But these American machines were mainly of the sort which could be easily harnessed to horse power (reapers, harvesters, etc.). In fact steam cable-ploughing did not make nearly as big an impact on American agriculture as did the light internal combustion tractor when this became available early in the twentieth century.

1829.

GRAND COMPETITION

OF

LOCOMOTIVES

ON THE

LIVERPOOL & MANCHESTER RAILWAY.

STIPULATIONS & CONDITIONS

On which the Directors of the Liverpool and Manchester Railway offer a Premium of £500 for the most Improved Locomotive Engine.

I.

The said Engine must "effectually consume its own smoke," according to the provisions of the Railway Act, 7th Geo. IV.

II.

The Engine, if it weighs Six Tons, must be capable of drawing after it, day by day, on a well-constructed Railway, on a level plane, a Train of Carriages of the gross weight of Twenty Tons, including the Tender and Water Tank, at the rate of Ten Miles per Hour, with a pressure of steam in the boiler not exceeding Fifty Pounds on the square inch.

III.

There must be Two Safety Valves, one of which must be completely out of the reach or control of the Engine-man, and neither of which must be fastened down while the Engine is working.

IV.

The Engine and Boiler must be supported on Springs, and rest on Six Wheels; and the height from the ground to the top of the Chimney must not exceed Fifteen Feet.

V.

The weight of the Machine, WITH ITS COMPLEMENT OF WATER in the Boiler, must, at most, not exceed Six Tons, and a Machine of less weight will be preferred if it draw AFTER it a PROPORTIONATE weight; and if the weight of the Engine, &c., do not exceed FIVE TONS, then the gross weight to be drawn need not exceed Fifteen Tons; and in that proportion for Machines of still smaller weight—provided that the Engine, &c., shall still be on six wheels, unless the weight (as above) be reduced to Four Tons and a Half, or under, in which case the Boiler, &c., may be placed on four wheels. And the Company shall be at liberty to put the Boiler, Fire Tube, Cylinders, &c., to the test of a pressure of water not exceeding 150 Pounds per square inch, without being answerable for any damage the Machine may receive in consequence.

VI.

There must be a Mercurial Gauge affixed to the Machine, with Index Rod, showing the Steam Pressure above 45 Pounds per square inch; and constructed to blow out a Pressure of 60 Pounds per inch.

VII.

The Engine to be delivered complete for trial, at the Liverpool end of the Railway, not later than the 1st of October next.

VIII.

The price of the Engine which may be accepted, not to exceed £550, delivered on the Railway; and any Engine not approved to be taken back by the Owner.

N.B.—The Railway Company will provide the ENGINE TENDER with a supply of Water and Fuel, for the experiment. The distance within the Rails is four feet eight inches and a half.

THE LOCOMOTIVE STEAM ENGINES,

WHICH COMPETED FOR THE PRIZE OF £500 OFFERED BY THE DIRECTORS OF THE LIVERPOOL AND MANCHESTER RAILWAY COMPANY.

DRAWN TO A SCALE ¼ INCH TO A FOOT.

THE "ROCKET" OF Mr ROBt STEPHENSON OF NEWCASTLE.

WHICH DRAWING A LOAD EQUIVALENT TO THREE TIMES ITS WEIGHT TRAVELLED AT THE RATE OF 12½ MILES AN HOUR, AND WITH A CARRIAGE & PASSENGERS AT THE RATE OF 24 MILES. COST PER MILE FOR FUEL ABOUT THREE HALFPENCE.

THE "NOVELTY" OF MESSrs BRAITHWAITE & ERRICSSON OF LONDON,

WHICH DRAWING A LOAD EQUIVALENT TO THREE TIMES ITS WEIGHT TRAVELLED AT THE RATE OF 20¾ MILES AN HOUR, AND WITH A CARRIAGE & PASSENGERS AT THE RATE OF 32 MILES. COST PER MILE FOR FUEL ABOUT ONE HALFPENNY.

THE "SANSPAREIL" OF Mr HACKWORTH OF DARLINGTON,

WHICH DRAWING A LOAD EQUIVALENT TO THREE TIMES ITS WEIGHT TRAVELLED AT THE RATE OF 12½ MILES AN HOUR. COST FOR FUEL PER MILE ABOUT TWO PENCE.

5: *Steam on Rails*

The most widespread form of the steam engine was probably the railway locomotive. Mill engines and public service engines were usually much larger and their social importance was almost certainly greater. But the railways went everywhere and affected everybody, so their impact on the public and the popular imagination was much greater.

The railway pre-dated the steam locomotive, as tramways had been used in mining and iron and stone-working districts in Britain during the eighteenth century. There were even earlier forerunners of tracked transport in the German mines of the sixteenth century. But it was the steam locomotive which changed the railway from being merely useful to mineral working into a general transport system. The first fully-fledged railway in the modern sense, with a time-table for goods and passengers, was the Liverpool and Manchester Railway, opened in 1830. The success of the *Rocket* at the Rainhill Trials in the previous year had made sure that locomotive steam power would be adopted for this line. The Trials were organized by the owners of the Liverpool and Manchester Railway to convince themselves, and their prospective customers, that the steam locomotive had sufficient power to climb the gradients on the route with a respectable load at a modest speed and without making too much smoke. The *Rocket* was the only competitor to meet these conditions successfully and so it won the £500 prize for the competition and was accepted for use on the new railway.

Rocket was built by George Stephenson (1781–1848), and

Opposite. The leading locomotives which completed in the Rainhill Trials of 1829, together with the conditions of entry.

Above. George Stephenson, the famous British railway locomotive engineer.

Below. An engraving of the famous Trials at Rainhill, won by George and Robert Stephenson's *Rocket*.

his son Robert (1803–59). Together with Newcomen, Watt, and Trevithick, the Stephensons are among the venerated pioneers of steam power.

Soon after Trevithick had fitted his compact high pressure engine to a road carriage in the streets of Camborne, he had set it to work on the Penydarren tramway in south Wales. This was on 21st February, 1804, and although it worked well the engine was not taken into regular service because it damaged the tramway plates. In the same year George Stephenson was appointed as engineer to Killingworth Colliery, near Newcastle. Here, he was able to observe the early attempts to introduce steam traction in the north-east coalfield. Fired with enthusiasm for steam locomotives, he began to build them himself. In 1821 he was appointed an engineer to the Stockton and Darlington Railway. The owners thought of this railway as a conventional mineral tramway. But Stephenson used its opening ceremony in 1825 to display his steam locomotive Locomotion No. 1, triumphantly hauling a loaded train of thirty-eight carriages at twelve m.p.h. Like most of the early steam locomotives, it had two vertical cylinders encased in the boiler to keep them hot, with the exhaust directed into the chimney to increase the blast in the fire-box. Today, Stephenson's engine can be seen on the platform of Darlington Station.

Despite the great achievements of the early locomotives built by Stephenson, Timothy Hackworth (1786–1850) and other ingenious engineers, railway owners were not convinced that they could always work as effectively as horse-

power or even a stationary steam engine. That is why the dramatic success of *Rocket* at the Rainhill Trials was so significant. It not only showed that the steam locomotive could do all that could reasonably be expected of it, but it also showed that in the hands of some brilliant engineers the steam engine could very rapidly be improved. It was a great technological advance on its predecessors and suggested some promising lines for future development. The young Robert Stephenson seems to have been mainly responsible for the important innovations, which included a multi-tubular boiler, an improved fire-box and cylinders placed diagonally on the outside of the boiler coupled directly to the driving wheels. It was a simple step to go on to place the cylinders under the boiler, driving the wheels through a cranked axle. This Robert Stephenson proceeded to do in *Planet,* which became the standard type locomotive used on the Liverpool and Manchester Railway, and was in a sense the prototype of later steam locomotives.

Above. George Stephenson's *Locomotive No. 1* of 1825 (on the right) on view at Darlington Station. The other engine, *Derwent*, of 1846, also operated on the Stockton & Darlington Railway.

Left. The *Planet*, built by Robert Stephenson, which became the standard type of locomotive used on the Liverpool and Manchester Railway.

47

1830 may thus be taken as marking the dawn of the railway era. Owners and potential investors remained cautious about the profitability of railways for a few more years, but they were soon convinced by the success of the Liverpool and Manchester Railway. It promoted the spectacular railway boom which gripped Britain for the next quarter of a century and which spread rapidly to Europe, North America, and the rest of the world. This boom brought about a great change in methods of transport and made easy travel possible for people who had never expected to move very far from their homes. The early railway promoters thought, like the tramway-builders before them, that the main use for railways would be the transport of minerals and other bulky commodities. The demand for passenger trains took them completely by surprise. However, they quickly developed their passenger services, and for several decades it was these which, in Britain at least, made the most profit.

The development of the British railway system was as confused as it was rapid. Erratic investment caused by ups

Below. Steam locomotive and passenger coaches on the Liverpool and Manchester Railway.

LONDON AND BIRMINGHAM RAILWAY.

HOURS OF DEPARTURE AND TIME TABLE, ON AND AFTER 1st OTCOBER, 1839.

DOWN TRAINS FROM LONDON.

TRAINS.	Departure from London.	Harrow.	Watford.	Boxmoor.	B. Hampstead.	Tring.	Arrival at Aylesbury fm London.	Leighton.	Bletchley.	Wolverton.	Roade.	Blisworth.	Weedon.	Crick.	Rugby.	Brandon.	Coventry.	Hampton.	Arrival at Birmingham.
*† MIXED, calling at 1st cl. Stns.	6 a.m.	—	6.45	—	—	7.25	—	7.50	—	8.15	—	8.50	9. 5	—	9.40	—	10.10	10.35	11½ a.m.
* MIXED	8 a.m.	8.30	8.50	9.10	9.20	9.35	10.15	10. 0	10.15	10.30	10.55	11. 5	11.25	11.45	12. 5	12.20	12.35	1. 0	2 p.m.
§†* FIRST, calling at Mail Stns.	8½ a.m.	—	—	—	—	10. 5	—	—	—	11. 0	—	11.35	11.50	—	—	—	12.50	1.15	2½ p.m.
§* MAIL	9½ a.m.	—	—	—	—	10.48	—	—	—	11.41	—	—	12.33	—	—	—	1.36	—	2½ p.m.
* MIXED, calling at 1st cl. Stns.	11 a.m.	—	11.45	—	—	12.25	—	12.50	—	1.15	—	1.50	2. 5	—	2.40	—	3.10	—	4½ p.m.
† MIXED, calling at 1st cl. Stns.	1 p.m.	—	1.45	—	—	2.25	—	2.50	—	3.15	—	3.50	4. 5	—	4.40	—	5.10	5.35	6½ p.m.
MIXED	2 p.m.	2.30	2.50	3.10	3.20	3.35	—	4. 0	4.15	4.30	4.55	5. 5	5.25	5.45	6. 5	6.20	6.35	7. 0	8 p.m.
MIXED, to Aylesbury	3 p.m.	3.30	3.50	4.10	4.20	4.35	5.15	—	—	—	—	—	—	—	—	—	—	—	—
MIXED, calling at 1st cl. Stas.	5 p.m.	—	5.45	—	—	6.25	—	6.50	—	7.15	7.40	7.50	8. 5	—	8.40	—	9.10	—	10½ p.m.
MIXED, to Wolverton	6 p.m.	6.30	6.50	7.10	7.20	7.35	8.15	8. 0	8.15	8.30	—	—	—	—	—	—	—	—	—
†§* MAIL, Mixed.	8½ p.m.	—	—	—	—	9.56	—	—	—	10.54	—	—	11.50	—	—	—	1. 0	—	2 a.m.

SUNDAY TRAINS.

TRAINS.	Departure from London.	Harrow.	Watford.	Boxmoor.	B. Hampstead.	Tring.	Arrival at Aylesbury fm London.	Leighton.	Bletchley.	Wolverton.	Roade.	Blisworth.	Weedon.	Crick.	Rugby.	Brandon.	Coventry.	Hampton.	Arrival at Birmingham.
* MIXED	8 a.m.	8.30	8.50	9.10	9.20	9.35	10.15	10. 0	10.15	10.30	10.55	11. 5	11.25	11.45	12. 5	12.20	12.35	1. 0	2 p.m.
§* MAIL	9½ a.m.	—	—	—	—	10.48	—	—	—	11.41	—	—	12.33	—	—	—	1.36	—	2½ p.m.
MIXED, to Wolverton	6 p.m.	6.30	6.50	7.10	7.20	7.35	8.15	8. 0	8.15	8.30	—	—	—	—	—	—	—	—	—
†§* MAIL, Mixed.	8½ p.m.	—	—	—	—	9.56	—	—	—	10.54	—	—	11.50	—	—	—	1. 0	—	2 a.m.

Above. Part of the London and Birmingham Railway timetable of 1839, showing the frequent services available.

and downs in business confidence and other disturbances resulted in an uneven rate of railway building. The first lines to be built after the Liverpool and Manchester Railway were the trunk routes to London. The London and Birmingham line, engineered by the Stephensons, came first, and the Great Western Railway linking Bristol with London following shortly after, engineered by the versatile genius Isambard Kingdom Brunel (1806–59). Once the main trunk lines had been staked out, there was a rush to build the subsidiary and feeder lines, and at this stage rivalry between the various companies became intense.

The next phase was soon reached. Railways started to compete directly with each other and negotiated to secure operating rights or direct control over lines to strengthen their bargaining position. By 1843, there were already two thousand miles of track in Britain, and by 1849 this had risen to five thousand miles. Every railway company needed an Act of Parliament before it could acquire land. Although certain obligations were imposed by law and there was always powerful opposition from other interests to be overcome, little attempt was made to control the railway development and to plan it sensibly. The final stage of railway expansion came when the companies merged into groups (often associated with particular trunk routes) and filled gaps in the railway network. They also built more tracks, especially where there was only a single line.

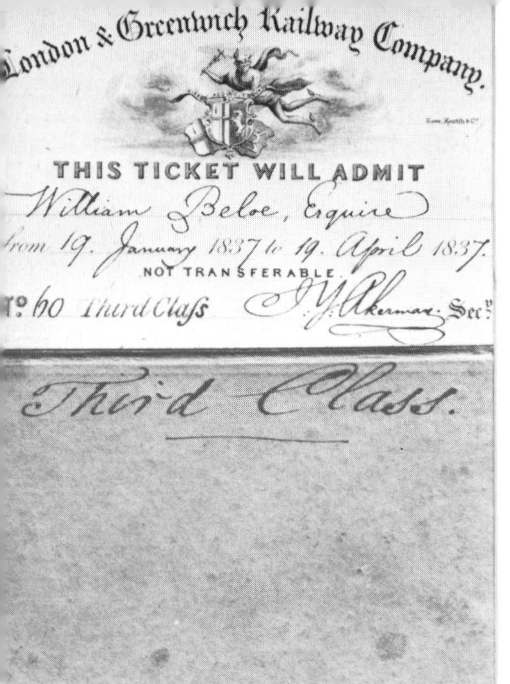

Above. One of the first railway season tickets issued by the London and Greenwich Railway to cater for the increasing number of commuters into London.

Below. A diagram of I. K. Brunel's "atmospheric railway" in S. Devon. His idea was to have the engine piston fitted into a continuous tube fixed between the rails.

On the thousands of miles of this railway system, the steam locomotive reigned supreme. It was only challenged when, in the 1840s, I. K. Brunel tried unsuccessfully to introduce an "atmospheric" method of propulsion on the South Devon Railway. He planned to have stationary engines creating a partial vacuum in a continuous tube between the rails, and with the trains attached to a piston in the tube.

Although Robert Stephenson had created the prototype steam locomotive in *Planet,* there were many variations. And as there was constant pressure on the railway companies to build more powerful and faster locomotives, the locomotive engineers could never relax, believing they had produced the ultimate form of steam engine. As late as the 1930s, with diesel power already replacing steam on American railroads and electric traction making progress in Britain on the southern suburban services, rivalry between the "Big Four" railway companies produced dramatic struggles for the speed record. This record was held, when the Second World War (1939–45) put a stop to such competition, by the L.N.E.R. (London and North-East Railway) Pacific Class *Mallard* which, in 1938, reached 126 m.p.h.

The effect of the railways and the steam locomotive on society is incalculable. Railways so changed life in Britain that it is unrealistic to ask what life would have been like without them. But it is possible at least to show some of the areas in which the railways made a great impact.

Firstly, the railways almost totally eclipsed other forms of inland transport. The Royal Mail was first carried by rail in 1837, and because people of all social classes from Queen Victoria downwards came to prefer the speed and comfort of rail transport to that available on the roads, the stage coaches were taken out of service and scrapped. The roads survived to make a spectacular come-back in the twentieth century, but the canals were not so fortunate.

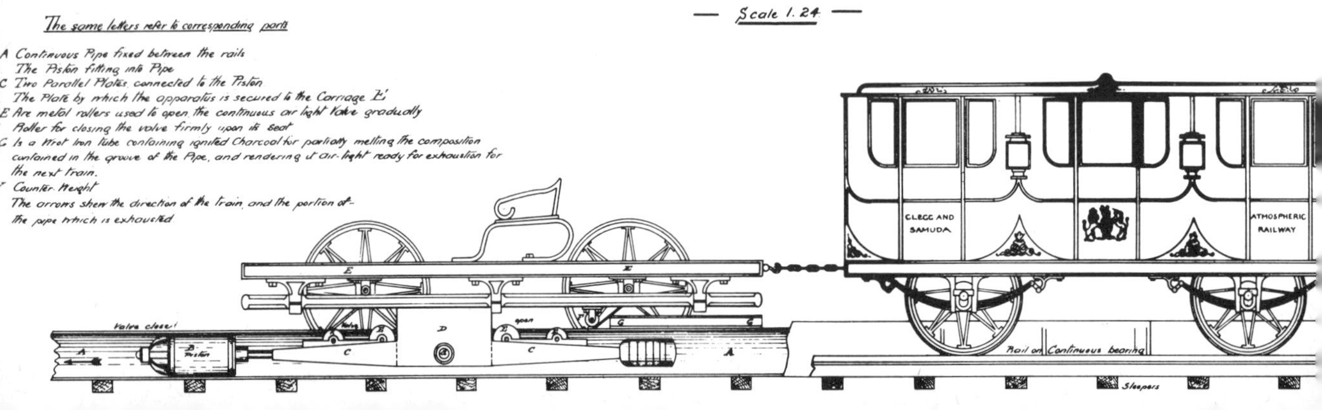

These had monopolized the transport of bulky goods by the 1830s, but the railways destroyed this monopoly and virtually eliminated the canals. Luckily, a few of the once flourishing canals have survived, but as a commercial transport system the British canals have little future. In other countries, where canals were sometimes built to carry large vessels, they remain in active use, but in Britain the narrow and winding canals could not compete with the railways. But it is worth remembering that the steam engine was used on many canals for pumping water to replace that lost in working the locks. A particularly interesting pair of such engines survive at Crofton on the Kennet and Avon Canal, and have recently been restored to full working order.

Secondly, rapid transport stimulated many forms of industrial activity which would have been impossible without it. Raw materials could be brought from further afield. Fuel could be acquired more cheaply. Markets could be reached more rapidly. Industrialists, commercial travellers, technical experts and skilled labour could move with unprecedented ease.

Thirdly, the railway construction and locomotive engineering industries boosted economic activity. The money made from the railway boom was unequalled by any previous enterprise, and its management needed far-reaching changes in the sharemarket and financial institutions of the country. The huge labour force of skilled

Above. The L.N.E.R. Pacific Class *Mallard* which held the speed record in 1938 with a speed of 126 m.p.h.

Below. A view of the inside of a Victorian railway coach, showing the luxury and comfort that made the railways a far more pleasant form of transport than the stage coach—at least for the first class passengers!

workmen and navvies employed in railway building needed to be efficiently organized and this led to the emergence of giant firms of contractors which could undertake such operations. The engineering industry, which had been born out of the needs of steam technology, grew very prosperous on the construction of locomotives and rolling stock. The large railway companies built large engineering workshops, and in places like Swindon and Crewe virtually created new towns.

Fourthly, the ease of passenger transport contributed to a marked growth in towns. The idea of commuting between home and work led to residential suburbs served by railways which by law had to run cheap workers' trains at the beginning and end of each day. The urban termini of the railways created a daily ebb and flow of workers of all

Below. Navvies working on the building of a railway in S. London in 1867. Large firms of contractors, employing vast numbers of labourers, emerged during the period of the railway boom.

social classes into and out of the towns. By supplying the urban community with fresh milk, fish, and other perishable foods the railways helped to enrich the quality of town life.

Finally, the railways played a big part in the "leisure revolution" of the second half of the nineteenth century. Railway bookstalls supplied a new type of reading, with journals like *Punch* and the *Illustrated London News* for the commuting public. Even more important, the excursion trains from towns to the seaside and other resorts meant places like Blackpool, Southend and Weston-super-Mare rapidly became holiday centres.

Taken together, these achievements of the British railway system caused a deep and lasting change in the way

Above. A railway bookstall in the late nineteenth century. Journals like *Punch*, the *Illustrated London News* and the *Daily Graphic* catered specifically for the reading interests of the commuting public.

Above. The Union Pacific Railroad in America. Most of the great transcontinental railways in America were built to carry freight goods rather than passenger traffic.

of life of the people of Britain. In other parts of the world, a similar set of influences was at work. But on the great transcontinental railways of America and on other trunk routes, economic and political factors were more important than social effects. Freight traffic dominated the American railroads, with bulk movement of grain to the seaports, meat to the canning factories and engineering supplies to farmers and backwoods miners. In Canada, the Canadian Pacific Railway, finished in 1885, had a powerful political objective. It linked the remote territories of the Pacific coast to the centre of government in Ottawa and the East. Similar political reasons inspired some of the railway

schemes in Latin America and Africa. The steam locomotive played an essential part in opening up the in-
teriors of continents to economic development, and in
strengthening the authority of distant governments in
these areas.

The steam locomotive gave way, in the twentieth century, to diesel and electric traction on most of the main
railways of the world. The railways have themselves been
superseded as the best method of inland transport by the
roads and internal combustion engine, and by the
development of airways. The outstanding service of steam
to land transport therefore lasted for about a hundred
years, from 1830 to the Second World War. But in this
period its service was not only outstanding—it was also in-
dispensable. The world would certainly be a poorer place
without the world-wide network of railways that steam
power made possible.

Left. A poster advertising the speed of
steam trains in the days before the coming of diesel and electric locomotives.

S.S. GREAT BRITAIN

6: *Steam Ships*

In the summer of 1970 the steam ship *Great Britain* returned to the Port of Bristol, to be restored in the dry dock which had been specially built for her construction, and from which she had been launched by H.R.H. Prince Albert (1819–61) on 19th July, 1843. This was a very remarkable event. The ship had been left to rust away in the Falkland Islands. It had seemed that nothing could be done to save her even though she was one of the most significant achievements of British marine engineering. Not only was the *Great Britain* the first large iron ship ever built, she was also the first large screw-propelled ship and one of the first steam ships to be designed specifically for crossing the oceans. Her career parallels the story of steam navigation, and it is entirely fitting that she should be restored as a permanent memorial to the steam ship.

The *S.S. Great Britain* was the creation of I. K. Brunel, who as well as being one of the pioneers of railway construction and a bridgebuilder of genius, turned his hand with great distinction and rare drama to shipbuilding. He built three major ships; the *Great Britain* was the second. His first ship was the *S.S. Great Western,* a wooden-hulled paddle steamer which was larger than any previously built when she was launched in 1837. Brunel thought of extending his broad gauge Great Western Railway from London to Bristol and on to New York via a regular passenger service in a steam ship specially built for the transatlantic route. Critics scoffed at the idea. Up till then steam ships had been small vessels—beginning in 1802 with William Symington's first practical paddle tug *Charlotte Dundas*

Opposite, above. The *S.S. Great Britain* (3270 tons) in dry dock at Bristol in 1975, 132 years after she was launched from the same dock by I. K. Brunel. *Below.* The paddle steamer *S.S. Great Western*, Brunel's first ship.

(intended for service on the Forth and Clyde Canal) and Robert Fulton's paddle steamer *Clermont,* built for river ferry duties in New York State in 1807. The early steam ship designers thought that a steamer on a long voyage would need to carry a prohibitive amount of coal. This meant they built small ferries, which could be re-fuelled easily, and discouraged experiments with larger ships. It was left to Brunel to show once and for all that the proportion of space required for fuel lessened as the size of the ship increased. So it was perfectly possible to build a larger steamer, equip it for long voyages and make a profit. This he did with the *Great Western.*

The *Great Western* was not the first steam ship to cross the Atlantic. The eastwards crossing had been made by the *Savannah* in 1819 using auxiliary steam power. The *Sirius* made the first sustained steam-powered westwards crossing in 1838. This ship had been specially equipped to make the crossing before the *Great Western* and succeeded in the attempt by only a few hours. But it was the *Great Western* which proved that steam ships could operate profitably on long voyages, and on the north Atlantic run in particular. Apart from her size (1320 tons compared with 703 tons for *Sirius*), the *Great Western* followed the nor-

Below. A sketch of the *Sirius* (703 tons) which made the first steam-powered westwards crossing of the Atlantic in 1838.

mal pattern of wooden ship construction, with side-lever steam engines driving paddle wheels. All the early steam ships used a vertical cylinder arrangement (except Symington's *Charlotte Dundas* which, strangely, had a horizontal cylinder). But as the traditional overhead beam affected the stability of a ship, marine engine designers usually replaced the beam by two low-level beams on either side of the engine. This was known as the "side-lever" design. Another design which became very popular with paddle steamers was the oscillating cylinder, which had two cylinders pivoted on trunnions so that their piston rods connected with the crankshaft at right angles to each other.

Having shown that profitable transatlantic business was possible for large steam ships, Brunel went on to exploit this market with his second ship, the *Great Britain*. This ship had, in its iron construction and screw propulsion, two important new features and, at 3270 tons, was by far the largest ship of her period. Although she was a striking technological success, the Great Western Steamship Company, which commissioned her, ran into considerable financial trouble and lost the valuable contract for the Royal Mail, which went to the Cunard Line. A serious

Below. William Symington's steam paddle tug *Charlotte Dundas*, built for service on the Forth and Clyde Canal.

accident in September 1846, when the *Great Britain* ran aground in north Ireland, bankrupted the company. But this accident did show the immense strength of the iron hull, for the ship was refloated the following year and spent many years on the run to Australia before being abandoned in the Falkland Islands in 1886.

Brunel then went on to build his third steam ship— the giant *Great Eastern* (18,915 tons). The keel was laid in 1854. She had a double iron hull divided by bulkheads into twelve watertight compartments, and was propelled by both paddles and a screw, each having its own set of engines. The paddle engines were four oscillating cylinders, and the screw engines four horizontal cylinders coupled directly to the propellor shaft. The ship was intended to take advantage of the developing Far Eastern market, but after great problems in building, launching and fitting out, her directors eventually sent her on her

Below. The *S.S. Great Eastern* (18,915 tons) in Millwall just before launching. She had a double iron hull, and was driven by paddles and a screw each having its own set of engines.

maiden voyage to New York in June 1860. Shortly afterwards she was chartered as a cable laying vessel and between 1865 and 1874 laid five transatlantic telegraph cables. One was even laid across the Indian Ocean from Suez to Bombay. Despite her great size (not exceeded until the launch of the *Lusitania* and *Mauretania* in 1906) the *Great Eastern* was a very manoeuvrable ship because of her mixed propulsion system. This made her exceptionally well equipped for the task of cable laying. But once this had been completed, there was no useful task left her and after being laid up for eleven years she was sent to be broken up on Merseyside in 1888.

By 1859 the steam ship was well established on the north Atlantic trade route and had taken over much of the coastal traffic around the shores of Europe. On the longer runs to South America and the Far East, steam ships were still not always profitable, but their numbers were increasing. For a while the magnificent last generation of sailing ships, the tea clippers, remained unchallenged. Two developments allowed the steam ship to capture this market from the sailing ship. One was the opening of the Suez Canal in Egypt in 1869, which considerably shortened the route to India and beyond. It created a short cut for steam-powered vessels which could not be used by the clippers because their speed depended on the trade winds in the open sea. Regular coaling stations situated on the Suez route further helped the steam ships, and so gradually after 1869 sail gave way to steam on the important eastern trade routes.

Below. The majestic sailing ships, the tea clippers, racing up the English Channel. With the opening of the Suez Canal in 1869, steam ships displaced sail on the eastern trade routes.

The other development that favoured steam was the striking improvement in the efficiency of the steam engine. The main reason for this was the introduction of compounding, by which the steam was passed at decreasing pressures through two cylinders. This was an innovation which affected all those who used steam power, and its industrial implications will be discussed in a later chapter. But, for the steam ship, compounding meant greater power from a more compact engine as well as an overall rise in efficiency.

The favourite form adopted for the compound engine, and for the "triples" and "quadruples" which followed it (with the steam being used in three or four cylinders respectively), was the inverted vertical arrangement. The cylinders were laid in line along the keel of the ship and drove downwards directly onto the propellor shaft. With twin screws, two sets of engines could be arranged in parallel. This lay-out could be fitted neatly within the frame of the hull, giving maximum stability to the ship. It became standard on large battleships and liners until the introduction of the steam turbine at the end of the nineteenth century.

The turbine also helped make the steam ship more efficient. It had been developed mainly for industrial purposes (which are discussed elsewhere). Charles Parsons (1854–1931), who invented the steam turbine in 1884, was quick to see its potential as a form of marine propulsion and his company built a small steam yacht, *Turbinia,* to test its value. This was the ship which raced at thirty-four knots among the warships of the British Navy at the Spithead Review in 1897, defying all attempts to overtake her. As a result Parsons showed the reluctant admiralties of Britain and Germany that they needed to re-equip their navies with turbine powered vessels. Turbines became obligatory for the two high performance liners then under consideration by the Cunard Line—the *Lusitania* and *Mauretania.* These had four screws each, two driven by high pressure turbines and two by low pressure turbines. The *Mauretania,* which averaged 27.4 knots on her trials, held the Atlantic Blue Ribbon (the prize for the fastest transatlantic crossing) until 1929. Her sister-ship was sunk in 1915 off the Irish coast by a German submarine.

Below. The steam turbine yacht *Turbinia*, which outraced every British warship at the Spithead Review in 1897, thus proving the superiority of turbine powered ships.

Companies like the White Star, which preferred a high standard of comfort to high speed, retained compound expansion engines for some time. The ill-fated White Star liner *Titanic,* which sank on her maiden voyage in April 1912 with great loss of life, was one of these, having two sets of triples exhausting into a low pressure turbine to drive her three screws.

Although steam propulsion has met with increasing competition from diesel engines in small and medium-size ships in the twentieth century, steam turbines are still used in many large ships. Virtually all steam ships, however, have switched to oil–firing for their boilers. Air transport has in the last few years taken much of the regular passenger service from the large luxury liners, and most of them are now used for cruising duties—and there are far fewer than there used to be.

The steam ship has done much to make the world a smaller place. Its regular and reliable service has enabled large numbers of passengers and an enormous volume of cargo to be transported to countries which had previously been inaccessible. It has helped many people to migrate westwards across the Atlantic and has made a great contribution to the settlement of Australia. It has brought essential food stuffs and raw materials to the advanced industrial nations, and delivered their manufactured goods to remote markets. In short, the steam ship has been one of the most faithful and fruitful servants of mankind.

Above. Most of the screw engine layout of the *S.S. Great Eastern*. The paddle wheels were driven by four oscillating cylinders, not shown in this picture.

Below. The boiler room of the *Queen Elizabeth II*, which is powered by oil-fired steam turbines like most large passenger liners still in service.

Above. A model of Richard Trevithick's original steam engine which he fitted to a road carriage at Camborne, Cornwall, in 1800.
Below. The steam-driven gun carriage, designed by Nicholas Cugnot for the French Army in 1769.

7: *Steam on the Road*

Nineteenth century society was probably wise to discourage steam traction on the roads and to insist on locomotive engines having their own track. Mainly it was interested parties such as the turnpike trusts and the railway companies, who were responsible for this. But it is interesting to think what might have happened if the internal combustion engine had, from the start, been subjected to the same harsh conditions, and if the notorious "Red Flag Act" had not been repealed in 1896. Despite discouragement, however, the steam engine did take to the roads.

The first experimental steam locomotives had been road vehicles. Nicholas Cugnot (1725–1804) had built a steam-driven gun carriage for the French army in 1769, but it had not been a success. William Murdock (1754–1839), the Boulton and Watt representative in Cornwall and an inventor in his own right, had fitted an engine to a road vehicle in 1785. James Watt, however, had discouraged the experiment as he was convinced that steam traction was impractical. He was probably right, since at a time when the standard steam engine was a low pressure vacuum condensing type, it was inevitably a large machine. However, the small but powerful high pressure engine built by Richard Trevithick immediately made it possible to build more compact locomotives. Trevithick was quick to realize this and constructed a steam carriage which successfully carried some boisterous passengers at Christmas 1801. Unfortunately, in the celebrations which followed, the engine was allowed to boil itself dry and set its shed on fire.

Above. The statue of Richard Trevithick, holding one of his early steam locomotives, outside the public library at Camborne.

Below. Sir Goldsworthy Gurney's steam carriage on the London to Bath road. For many years, high tolls and the "Red Flag Act" of 1865 discouraged this form of transport.

Trevithick displayed another steam carriage in London in 1803, but then switched his attention to the railway locomotive and the steam road carriage was neglected. At this time, when the idea of the steam locomotive was a novelty, poor roads and the hostility of other road users effectively discouraged further experiments, but not for long.

With the steady improvement of the railway locomotive, and mounting enthusiasm and excitement over its possibilities, people started experimenting with road steam carriages once again. One of the *Rocket*'s competitors in the Rainhill Trials of 1829 was the steam carriage *Novelty,* built by John Braithwaite (1797–1870) and John Ericsson (1803–89) and adapted for use on roads. Although this carriage was not very powerful, it had a good compact design and could well have been developed as a road vehicle. Sir Goldsworthy Gurney (1793–1895) built an even better steam carriage in 1831 that looked like an elongated stage coach. It ran a regular service between Gloucester and Cheltenham, completing the nine mile journey in forty-five minutes. Other vehicles were just as successful in other parts of the country and it was clear that most of the technical problems had been solved. The widespread use of this form of steam locomotive was only prevented by the turnpike trusts who imposed high tolls, and by the "Red Flag Act" of 1865. This Act limited all mechanically propelled road vehicles to a speed of four m.p.h. and made them carry a crew of two with a third man to go ahead of the coach waving a red flag to warn other road users.

Even this law could not completely stop the development of steam on roads. Some elegant, covered carriages were made in the 1870s and in the following decades steam tricycles were imported into Britain from France, where they had been developed by Leon Serpollet (1858–1907). Yet the only form of steam locomotion to escape the crippling restrictions was the heavy traction engine. This seems to have evolved along with the steam ploughing engine which was introduced, as we have seen, in the mid-nineteenth century. It was a simple step to make a large portable engine self-moving and use it for many jobs on the farm. It was designed with the cylinder mounted above at the front of the boiler and was driven by gear transmission and controlled by a steering wheel. In the 1870s it pulled all sorts of things from ploughs to haycarts. The traction engine also became the mainstay of British fairgrounds, moving the equipment from site to site and providing power for the roundabouts. As a transport vehicle for breweries and other heavy-duty haulage firms, it was so efficient and reliable that it competed for a long time with the petrol lorry. It was also adapted to become the steam roller.

The first steam roller appeared in Britain in 1866, when Edward Aveling (1851–98) and Charles Porter (1826–1910) introduced a design from France which became very popular in road works for flattening and pressing down the

Below. One of the longest-lasting applications of the steam traction engine—the steam lorry, which was widely used as a transport vehicle for breweries.

Above. Carefully preserved steam engines being displayed by their owners at a modern "steam traction rally".

surface of macadamized roads, and later for rolling asphalt.

In all these forms, steam traction gave valuable service on the roads and several engineering firms were famous for their strong and powerful machines. Many of these engines are now privately owned by people who cherish them and display them at "steam traction rallies."

The easing of restrictions on mechanized road transport in Britain in 1896 gave an enormous boost to the new petrol motor cars. Although it also helped the steam carriage, it came too late for steam to make up lost ground and to compete effectively with the internal combustion engine. Nevertheless, some very successful steam cars were produced before the First World War (1914–18), after which the mass produced motor car took over. One of these steam cars was the Stanley which was developed in America. This vehicle had double cylinders, a petrol burner, and produced steam rapidly. It could run at high speeds and had a smooth action, with no gear changing. As modern industrial civilization is faced with an energy crisis and environmental pollution, steam cars may once again come into their own on the roads.

8: *Advanced Industrial Uses*

Steam technology was well represented in the Great Exhibition held at the Crystal Palace in 1851. The Palace itself, designed by Sir Joseph Paxton (1801–65), was a superb expression of new building techniques. With its pre-fabricated iron parts and thousands of standardized panes of glass, it was built with astonishing speed. Inside were displayed the products of the first fully industrialized society. It was seen as "the workshop of the world" because the products of British industry were then finding their way into the markets of every country. The Great Exhibition showed the self-confidence of a nation which knew that it had achieved much, and did not doubt that these achievements would continue. Behind both the achievement and the self-confidence was a commitment to the steam engine. It was clear that Britain had excelled largely because of its superiority in steam power and, as long as this superiority was not challenged, it seemed as if British leadership would be maintained.

It did not work out as simply as this. Firstly, British leadership in industrialization in the mid-nineteenth century could not be taken for granted. Certainly, Britain had led the way in increasing output with mechanization and new types of industrial and commercial organization. But once Britain had shown the way other nations were bound to follow, and many of them had much greater resources both in population and raw materials. So it was only a matter of time before Britain was overtaken. By 1913 iron and steel production in Germany and the U.S.A. was greater than that in Britain and many other nations were

Above. Electricity reaches the home. Electricity was to take over many of the tasks previously done by the steam engine, most notably in industry.

catching up.

Secondly, as the nineteenth century wore on it became apparent that steam technology was not going to remain unchallenged. As the steam engineers got better at turning heat into work, the whole science of heat and work—thermodynamics—was advancing. A direct result of this was a rapid improvement in a new type of engine—the internal combustion engine. Also, the science of electricity was beginning to show practical results. Although electric power was largely generated by steam engines, the abundant supply of electricity challenged the use of the steam engine in almost all its major forms.

Thirdly, Britain was even losing ground in steam technology itself in the second half of the nineteenth century. British ingenuity and craftsmanship were not at fault, but as the steam engine approached its maximum theoretical efficiency the scope for further improvements became less. The competition, too, of Swiss, French, Belgian, Italian, German, and American engineers, (many of whom had learnt their trade on British machines and in British workshops) became more intense. For example, the great improvement in steam valves invented by the American engineer George Corliss (1817–88) was widely adopted in Britain and elsewhere. Steam technology thus became much more of an international enterprise than it had been before, and the leading role of British engineers diminished correspondingly.

Britain found herself in a more highly competitive international situation in 1900 than in 1851. The new conditions called for economic and social changes which were not easy to make. In particular, the heavy financial investment in steam plant in the major sectors of the British economy may have affected the flexibility of these industries. But it is possible that other factors contributed to the inability of industry to adjust to the new situation. However, it was the leading sectors of the economy, which were dependent on steam, that were hardest hit by the inter-war depression in the twentieth century. As a result reciprocating steam engineering suffered greatly. When the depression hit the cotton textile industry, the coal mining industry and the iron and steel industry, dozens of steam engine manufacturing firms were forced out of business. In this crisis the reciprocating steam engine suffered a severe setback from which it has never recovered. Nevertheless it is worth remembering the high degree of mechanical perfection achieved by these large engines before 1900.

Much of the improvement in steam engine efficiency in the second half of the nineteenth century was due to the successful development of means of "compounding" engines. "Compounding" means using the expansive power of the steam at two or more stages in progressively larger cylinders. This technique dated back to the beginning of the century, when Arthur Woolf (1776–1837) patented, in 1803, a method of applying a high pressure alongside a low pressure cylinder in a traditional beam engine. But difficulties in getting the best ratio between the

Below. The British section at the Paris exhibition of 1878 displays the products of British technology at a time when it was beginning to lose its leadership in steam technology to foreign competition.

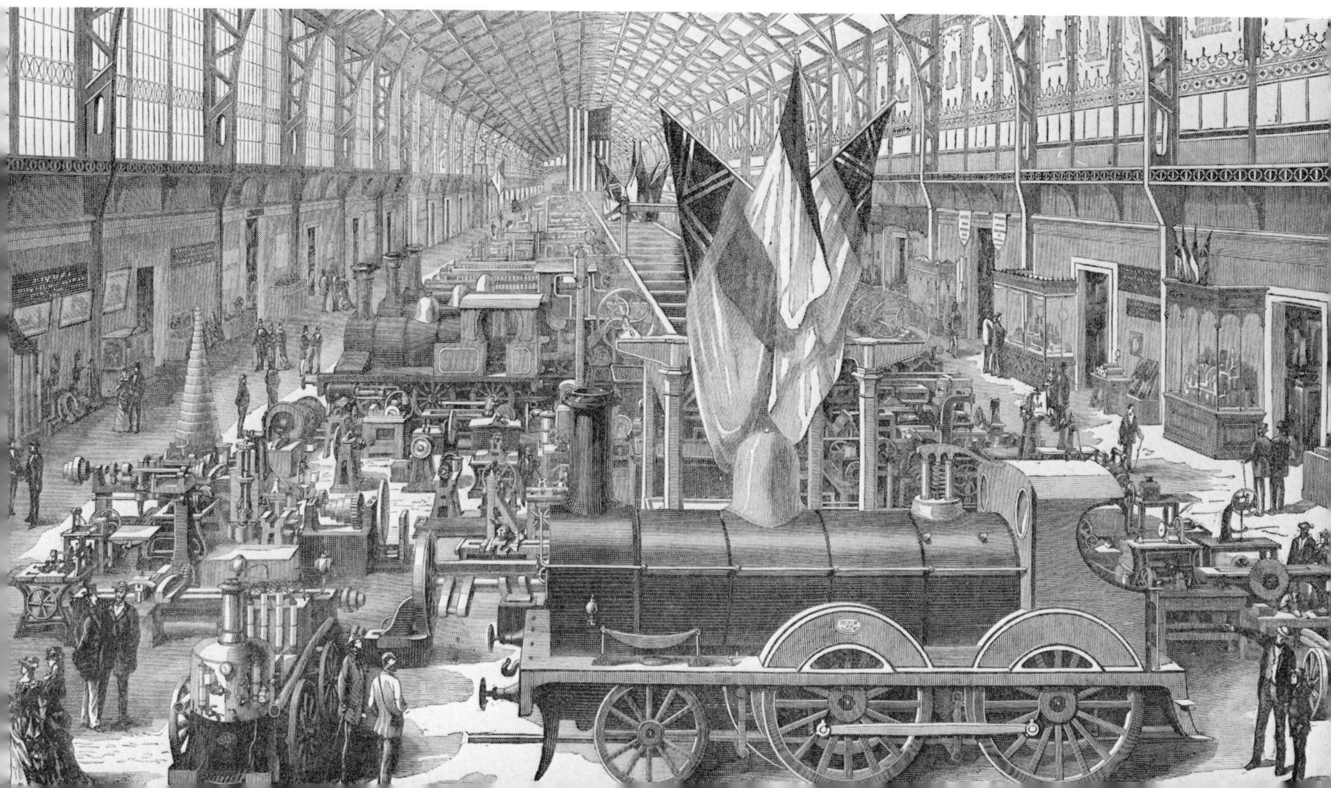

Above. Arthur Woolf's "compound" engine. Both the high pressure and the low pressure cylinders are at the far end of the beam.

Below. James Nasmyth's steam hammer at work. The inverted vertical steam engine was a simple development from this.

volumes of the two cylinders (and other constructional problems) at first made this type of engine unpopular in Cornwall. It had most chance of success in Cornwall because it was there that the efficiency of high pressure engines was most closely studied. Woolf's method of compounding was later much improved and widely used. It was only in 1845, when William McNaught (1813–81) introduced his system of compounding with the high pressure cylinder placed on the opposite side of the beam from the low pressure cylinder, that compounding became generally popular. Many beam engines were then "McNaughted" by adding another cylinder to give greater power and efficiency.

Another way of compounding a low pressure beam engine was to apply a high pressure horizontal engine to the same crankshaft and to feed the exhaust steam from the high to the low. Horizontal engines could themselves be compounded by adding another cylinder behind the first (a "tandem compound"), or parallel to it on the opposite side of the flywheel (a "cross compound"). By increasing the steam pressure it became possible to add a third cylinder ("triple expansion") and even a fourth ("quadruple expansion"), but the last was rare in industrial applications. At last steam technology had found a way of fully exploiting the expansive power of steam.

In the second half of the nineteenth century, the beam engine lost ground to the horizontal compound engines and to the inverted vertical engine derived from James Nasmyth's (1808–90) very successful steam hammer. In Nasmyth's engine piston rods drove downwards from a vertical cylinder directly onto the crankshaft. We have

already met this type of engine in marine engineering, but it also had many industrial uses.

Large steam engines like these were installed to drive mills, factories and workshops in the second half of the nineteenth century. These power plants encouraged the development of very large mills in the textile industries. In the manufacture of cotton goods especially they led towards increased specialization. Large mills, therefore, concentrated solely on spinning, or weaving, or other processes. In large textile mills of this sort the engine was usually installed in a house at one end of the main building, driving by ropes off the broad flywheel to shafts on the several floors of the mill. There were, however, important variants on this plan, such as the older method of driving by gearing off the flywheel to vertical shafts, which transmitted the power to horizontal shafting on each floor level. Such mills were common in south Lancashire and the West Riding of Yorkshire, and in other places where the textile industry flourished. In this way the steam engine performed one of its outstanding industrial services because the development of textile industries played a leading role in the industrialization of most Western nations.

Large steam engines were also installed in collieries for winding and other duties. But fuel costs were not so important in the mining industry and compounding and other sophisticated devices were not as attractive. It was quite common for colliery steam engines to do away with condensers and to exhaust steam directly into the air. But in maintaining the increase in national coal production, these machines played a vital part in the British economy

Below. The type of large horizontal mill engine that appeared in the second half of the nineteenth century. The flywheel, which transmits power by ropes to the driving shafts on the various floors of the mill, can be seen in between the high and low pressure cylinders.

Above. Few reciprocating steam engines remain active in industrial services. This steam winder at Penallta colliery, Glamorgan, was at work until the 1950s.

until the middle of the twentieth century. Then the nationalization and reorganization of the coal industry resulted in a greater use of electric winders. Very few steam engines are left in the modern coal industry. Those that are still in use are among the last of the reciprocating steam engines. Thus, for over two hundred and fifty years, the steam engine has served as "the miner's friend."

The British iron and steel industry is also moving away from reciprocating steam power. Most of the giant blowing and rolling engines installed in the nineteenth century (and, in some cases, later) are being taken out of commission. Continuous rolling processes in the heavy steel and tinplate industry are better suited to electric power. The specialized engineering industries like those making automobiles and electrical goods, which have grown so rapidly in the twentieth century, have relied on electric power almost from the start. The advantages of having machine tools equipped with individual electric drives has meant that electric power is now used throughout the engineering industry. The only places where steam plant has some advantages are in certain process industries where the exhaust steam is useful for purposes other than that of providing power. And so some paper, tobacco, food and drink processes, which need steam for heat, still use reciprocating steam engines.

The steam engines which remain in industry are mainly the advanced models, which achieved standards of efficiency hard to beat. Amongst these are the high speed enclosed engines with forced lubrication which developed at the end of the nineteenth century to act as electricity generators. Another late form of the reciprocating steam engine was the uniflow, so named because the steam flowed through the engine, leaving by ports in the middle of the cylinder, without having to reverse its movement.

The increase in efficiency of the reciprocating steam engine was partly due to the deeper theoretical understanding gained by scientists and engineers from the study of the performance of earlier engines; and especially the high pressure engines first successfully used in Cornish mines. We have already noticed that this advance in theoretical knowledge helped an understanding of the laws of thermodynamics and the development of the internal combustion engine. But it should be realized that it contributed also to the improvement of steam technology. The British found themselves at a disadvantage here. The invention and early development of the steam engine had not required a high level of theoretical knowledge, although (as we have seen) some understanding of the nature of a vacuum and the weight of the atmosphere was essential. Later, when the mechanical engineer had to understand the laws of thermodynamics, the lack of adequate elementary and technical education in Britain became obvious. Countries such as France, Germany and America, which had better methods of technical instruction, were therefore able to overtake British achievements in steam technology. But it is possible to see the steam engine as contributing to the development of education in Britain—one of the reasons for the introduction of general elementary and technical education in the second half of the nineteenth century was to improve the competitiveness of the nation in the field of scientific and technological theory.

9: *Public Services*

Below. Sewage spewing into the river Fleet, London, in 1844. The health hazard caused by such pollution made the Victorian civic authorities start massive sewage disposal schemes in which the steam engine played a crucial role.

The growth of cities has been one of the most notable and striking aspects of industrialization. The British National Census of 1851 revealed that, for the first time, most people were town-dwellers. And most of the towns which showed spectacular increases in population had been almost non-existent only two centuries before. The growth of towns had been accelerated by the steam engine, because the introduction of rotative action encouraged businesses (hitherto confined to the countryside by dependence on water power) to move towards the centres of population. Here, large steam-powered mills, factories and forges had been built, causing even more people in the surrounding districts to move into the towns in search of work. By 1850, this largely unplanned and uncontrolled expansion of towns was creating problems of law and order and disease. National leaders and administrators were worried. Their anxiety changed to alarm as cholera epidemics hit the towns of Britain in the 1830s, 40s and 50s. Slowly at first, but with increasing confidence and determination, the problems of town life in Victorian Britain were tackled. There was a tremendous financial investment in new water supplies, sewage disposal, urban transport services and municipal gas supplies. In all of these public service industries the steam engine played a vital role and, until very recently, this is where the largest number of steam engines have remained at work.

The use of steam engines in water works goes back to the time of Thomas Newcomen, when an atmospheric engine was installed at the York Buildings waterworks in London in 1726. Despite noisy opposition from contemporary pamphleteers, (one of whom feared that the population of the metropolis would be poisoned "through a long proboscis, something like an elephant's trunk"), this engine seems to have functioned well until 1731. It was replaced by a larger Newcomen engine, and steam power was soon a vital feature of the water supply of a metropolitan area. Paris had also experimented with steam engines for this purpose in the 1720s; and one of the first steam engines in America was set up in New York in 1776 to supply that city with water.

As cities grew in size and the public health hazard increased, many people realized that a supply of pure drinking water would be a great help in preventing disease. Even before the Public Health Act of 1848, many British towns and cities had begun to improve their water supplies by setting up companies to build reservoirs and purification plants, to sink wells and to pump water to the urban population. In the second half of the nineteenth century a large number of such civil waterworks were built. Most of them relied heavily on steam engines for pumping water from wells or springs into the primary reservoirs, and from there to the purification station and distribution reservoir. Large beam engines were usually installed as the emphasis was on steady pressure rather than speed. Cornish engines were frequently adapted for this purpose and, towards the end of the nineteenth century, water pumping engines were often compounded on the Woolf pattern (with high and low pressure cylinders side by side) for greater efficiency. Waterworks later made use of many other forms of steam engine for pumping duties, including some inverted vertical triples, such as those which replaced the beam engines in some of the Bristol Waterworks pumping stations. Most urban waterworks now depend entirely on electrical pumping apparatus. But a few large stations remain steam-powered—for example, that of the Metropolitan Water Board at Kempton Park.

In addition to supplying pure water, Victorian civic authorities felt obliged to remove foul water and therefore introduced comprehensive sewage disposal schemes. In large cities, such as London, such a task called for great civil engineering skill. The Metropolitan Board of Works

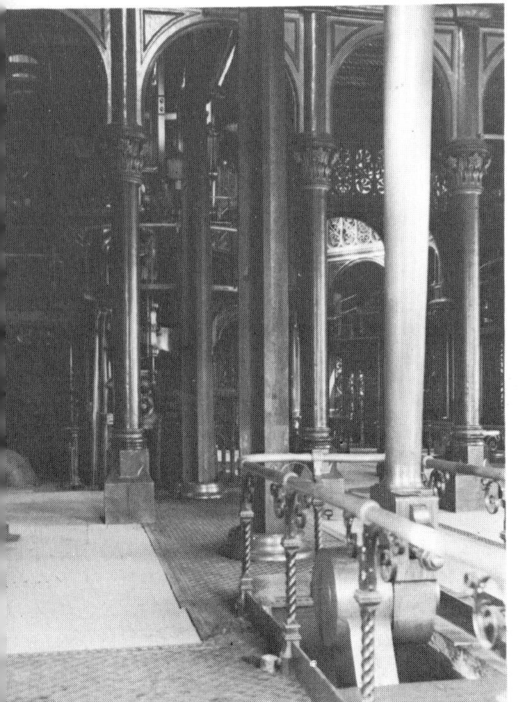

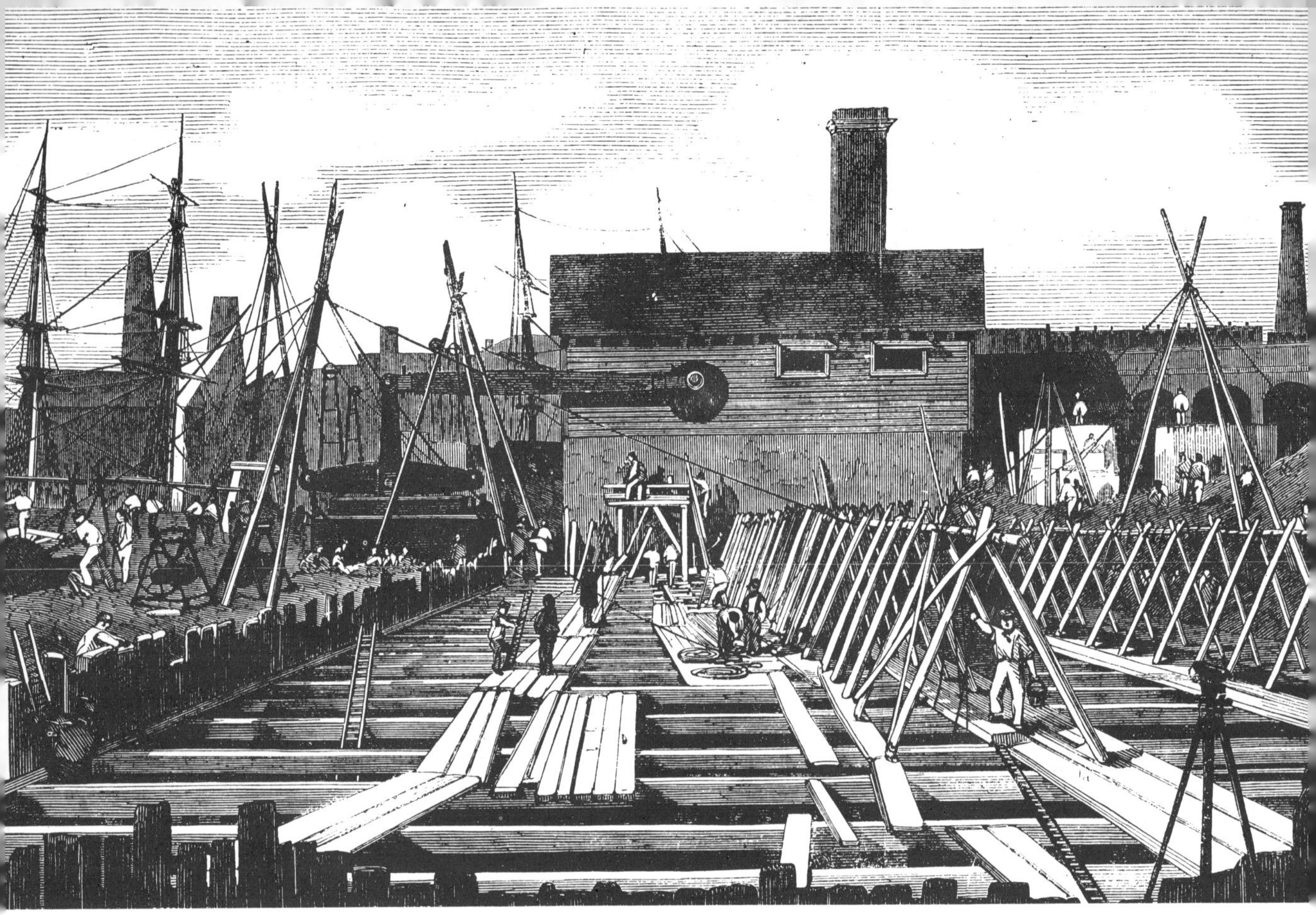

Above. The temporary steam pumping station at Deptford Creek installed during the construction of the London sewage disposal scheme. *Below.* A corner of the ornate engine house at Crossness pumping station.

in London were lucky to employ Sir Joseph Bazalgette (1819–90) as their engineer. It took Bazalgette twenty years from 1855 to complete his scheme, and involved building trunk sewers along both the north and south sides of the river Thames. The outfalls were situated well down-river from the metropolis. The scheme made use of four steam pumping stations with twenty large beam engines in all to keep the foul water moving. Those at Crossness Pumping Station on the southern outfall are a particularly distinguished set of beam engines. They have been preserved, athough they are no longer in use.

Other towns did not need such large sewage disposal works. But if the town was inland, the problems of disposing of waste and of returning relatively clean water to the rivers were great enough. In almost every town, at least one steam engine was installed for this purpose, and quite a number of these have survived to become tourist attractions, such as those at Shrewsbury (Coleham) and Cambridge (Cheddars Lane).

Another public service which used steam power extensively was the town gas supply. From the beginning, the gas industry was closely connected with steam power—William Murdock, the Boulton and Watt agent in Cornwall, had been the first person successfully to obtain gas from coal and burn it for illumination. Coal gas was first adopted on a large scale when it was used to illuminate Boulton's Soho factory in 1798. It took two more decades for the new fuel to be recognized as a source of lighting. But after that it was quickly used to illuminate street lights and the homes of nineteenth century Britain.

Steam power was necessary in town gas works for two purposes. Firstly, to exhaust gas from the retorts and to propel it through the various purifying processes. And secondly, to boost it into the gas holder to provide enough pressure to distribute it to the consumer. The engines required were generally quite small, although a large gas works could have a dozen or more steam engines kept in reserve. They gave excellent service in gas works until very recently. Lately, however, the substitution of oil-based and natural gas for coal gas, and the introduction of high pressure gas mains, has made most of the gas works obsolete. Their equipment, including most of their steam engines, has been scrapped.

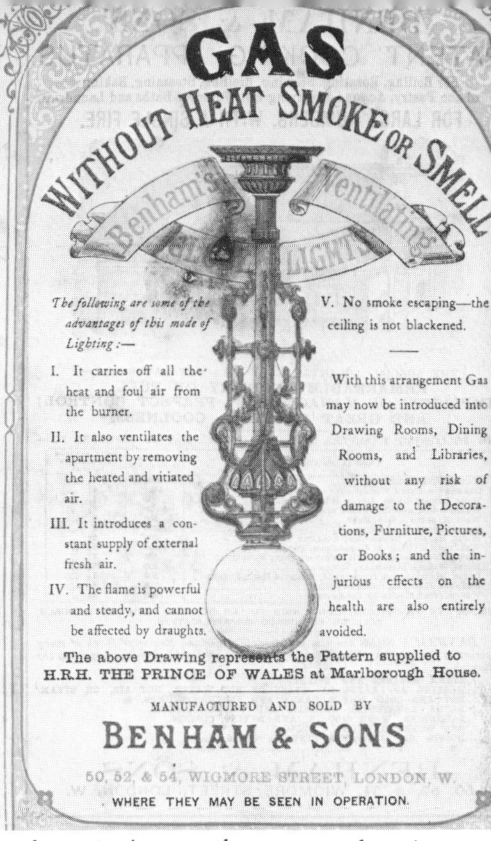

Above. A nineteenth century advertisement for gas lighting. *Below*. A painting of William Murdock lighting the gas lamps he installed in his house at Redruth, Cornwall, in the 1790s.

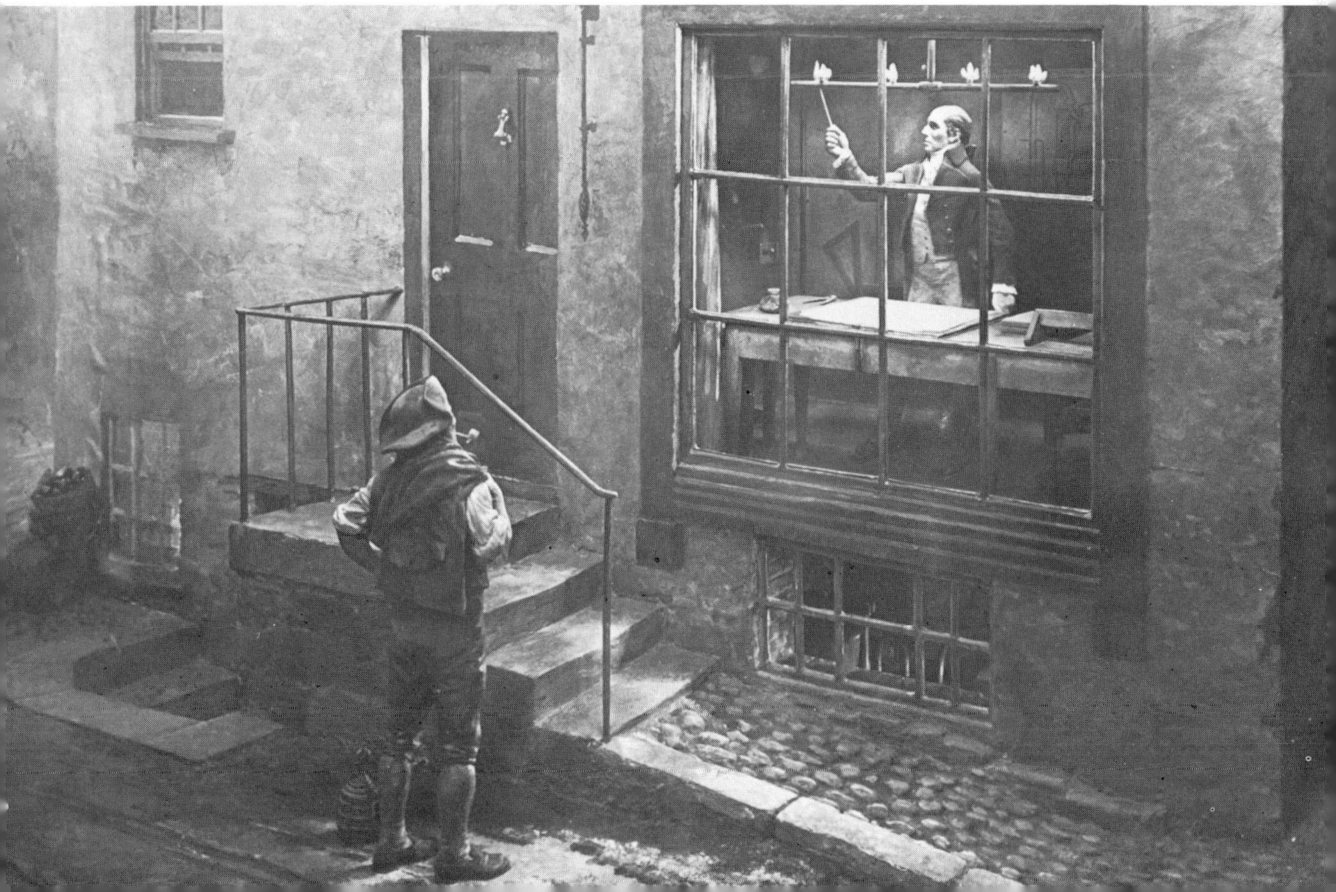

The role of the steam engine in providing a suburban railway transport service has already been mentioned. Apart from this (and some unsuccessful experiments with steam trams) the main function of steam power in city transport was in generating power for the electric trams. This, however, was only one aspect of a much more important service—that of generating electricity for the urban consumer. There were great technical difficulties in supplying electricity on a commercial basis. Although most of the problems were almost solved by the 1880s, public electricity supplies had only just begun by then. But the electric tram and other forms of electric traction made the public more familiar with electricity, and after 1900 it spread very rapidly to become the main source of energy for modern society.

The spectacular growth of electric supply would not have been possible without some important improvements in the steam engine. To get maximum efficiency from their dynamos, the electrical engineers needed engines which were powerful, compact and capable of a high speed of rotation. The ordinary reciprocating engine did not completely fulfil these needs, and a great deal of inventive thought went into making the necessary improvements. All the problems were eventually solved by the invention and development of the steam turbine. But before the turbine became generally available some very remarkable advances were made in the reciprocating engine to equip it for generating electricity. In 1884 (the same year that Parsons took out his first turbine patent) P. W. Willans introduced his central valve engine, (the valves were inside the hollow piston rod) which was capable of high speeds and was widely adopted in the early British power stations. The next step was to enclose all the working parts of a reciprocating engine so it could use forced lubrication. This led to the development of very fast, compact, engines such as those produced in thousands by the Birmingham firm of Belliss and Morcom. These were used for small-scale power generation as well as many traditional industrial purposes, and this sort of engine may be seen as the highest stage of development of the reciprocating steam engine. However, for large-scale electricity generation in the twentieth century, no reciprocating engine was

able to compete with the steam turbine, which made possible a great increase in efficiency and capacity. The outstanding service of the steam turbine has been in the generation of electric power. It is in this form that steam still makes a great contribution to the public services as well as to the industrial manufacturer and the domestic consumer.

It is worth noting, incidentally, that the great increases in the size and efficiency of steam engines was made possible by an improvement in steam boilers. The early boilers had been little more than large kettles, but the introduction of high pressure steam had meant that boilers had to be capable of delivering great volumes of steam at ever increasing pressures. The continuing development of steam boilers was thus an important factor in the success of steam technology in the nineteenth century. It remains of great significance in modern power stations.

Above. One of P. W. Willans' engines which was used for generating electricity. This type was installed in many British power stations.

10: *Modern Power Generation*

The modern atomic power station seems to have no connection with the primitive beam engines which lifted water out of the eighteenth century coal mines. But the link between them is in fact continuous and unbroken. In both cases the prime mover is a steam engine. One was a wasteful machine of a few horsepower, making steam by burning coal and condensing it in a cylinder to get a power stroke from atmospheric pressure. The other uses the heat from an atomic pile to make steam, which is then passed through huge efficient turbines of thousands of horsepower, to generate electricity. But both rely entirely on the elastic quality of steam—its ability to expand and to be condensed. In some form or other, therefore, the steam engine has made a continuing and crucial contribution to our industrial society.

Most steam power today is converted into electric current, and the steam turbine has made it possible to keep pace with the colossal growth in demand for electricity. It achieved (in a compact form and with a minimum number of moving parts) a direct conversion of energy from the expansion of steam into rotary motion. The invention of a way of using the expansive power of steam had taken a long time, first involving familiarity with the vacuum and atmospheric pressure. The introduction of direct rotary action was just as circuitous. Despite the early ideas of a primitive steam reaction turbine in the shape of Hero's aeolipile, engineers had failed to discover a workable method of harnessing rotary action, even though such distinguished inventors as Watt and Trevithick had tackled the problem. However, great improvements were made in the performance of water wheels in the nineteenth century, until finally the wheel was enclosed in a casing, with specially shaped buckets which were turned by a jet of water under pressure. This was the water turbine, and it replaced many of the surviving industrial water wheels. It continues to be used to produce hydro-electric power in mountainous areas where there is a reliable supply of water.

The steam turbine was a development from the water turbine but, unlike the water turbine, the energy in expanding steam is not completely used up in one wheel. The problem was to convert *all* the energy into rotary motion. This difficulty was solved by a brilliant young British engineer, the Hon. Charles Parsons (1854–1931), who patented the first steam turbine in 1884. Parsons' solution was to pass the steam through the blades of a series of wheels fixed to the same shaft or motor. Designed specifically to meet the need of the electric dynamo for a speed of 1200 r.p.m. (revolutions per minute)—which was too fast for the conventional reciprocating engine—Parsons' engine achieved a speed of over 12,000 r.p.m. He immediately started to design a dynamo that could be used with the turbine.

Below. Charles Parsons' original steam turbine, patented in 1884, which achieved a speed of 12,000 r.p.m. by passing steam through the blades of the wheels fixed to the shaft of the motor.

Having also overcome the other mechanical problems of operating at high speeds, the resulting turbo-generator unit worked very successfully. But as its steam consumption was initially high, potential users regarded it with caution. However, a Parsons unit was first installed in a public power station in 1888, when a seventy-five kw. (kilowatt) turbo-alternator was adopted by the Newcastle and District Electric Lighting Company. By the following year, when Parsons quarrelled with his business partners and temporarily lost control of his patents, some three hundred turbines had been supplied to a wide variety of users. As a result of Parsons' dispute over his patents, he switched his attention to the development of a radial flow turbine (with the steam moving outwards from the centre instead of the more "axial" arrangement in which it moved parallel to the motor.) He installed a hundred kw. unit of this type for the Cambridge Electric Lighting Company in 1891. This was the first condensing turbine, as Parsons had included a condenser to create a vacuum at the exhaust end, thus achieving even greater efficiency.

Parsons recovered full control over his patents in 1893, and shortly afterwards his company supplied two turbo-alternators of a thousand kw. to the German city of Elberfield. The patents were extended to 1903, but by that time other manufacturers had realized the tremendous

Below. 1,500 Kw. Parsons turbines in the main hall of a power station in 1907.

potential of the steam turbine. They were either producing Parsons' designs under licence or were developing their own ideas. The Swedish engineer Carl De Laval (1845–1913), for instance, had brought out a very successful design for an impulse turbine in which the steam was admitted through an expanding nozzle into buckets on the edge of a wheel. It reached speeds of up to 40,000 r.p.m. Charles G. Curtis (1860–1936) made significant improvements in steam turbine technology in America, including a British patent for pressure-compounding in 1896. The Swedish inventor Birger Ljungstrom made another important step forwards in 1907 with the successful development of double rotation: that is, instead of the blades on the rotor wheels being separated by static sets of blades on the inside of the casing, they were mounted on a second rotor moving in the opposite direction, thus adding considerable power. A thousand kw. turbo-alternator of this design achieved an efficiency of 77% in 1912 and was installed at the Willesden Power Station in England.

The First World War temporarily halted the advance of the steam turbine and delayed the phasing out of the reciprocating engine. Already, however, the pattern of large turbine installations had been set by Parsons' contract for a 25,000 kw. machine for the Commonwealth Edison Company Fisk Street Power Station, Chicago, completed in 1914. In 1918, the trend towards ever larger turbine units was resumed. This was particularly evident in America, where the Hell Gate Station in New York took a 168,000 kw. reaction set in 1928. And the State Line Generating Station, Illinois, installed a 208,000 kw. impulse set in 1929. In Britain, too, there was a move towards replacing the reciprocating engines by turbines for electricity generation, and towards larger and larger capacities. The introduction of the National Electricity Grid in the 1920s greatly speeded up this development, as large generating stations proved to be more economical than small ones and so produced cheaper electricity. As an electricity generating machine the turbine had enormous advantages because, despite improvements in the high speed performance of reciprocating engines, they could not match turbine speeds. Power for power they took up more space and caused more vibration. Thus the steam turbine came to enjoy a monopoly in this field, except in those areas where the water turbine was more suitable for the production of hydro-electricity. What has changed has been the means of raising steam for the turbines. Oil firing competed first with coal-burning and then with the development of atomic power stations. But as the twentieth century could well end in an acute "energy crisis," it is likely that coal fired power stations will continue to be used to generate electricity.

From 1900, then, the steam turbine rapidly became the main source for energy supplies. The extension of the National Grid, begun in Britain in the 1920s, has provided electricity for nearly everybody. In effect the steam engine now provides power for almost every home in the country and for a large part of its industry and transport system. So far as most people today are concerned, the power for lighting, heating, cooking, washing, cleaning, television and a variety of household gadgets, is based on that faithful work-horse of modern civilization—the steam engine.

Below. A modern 2,000 Mw. coal-fired power station near Retford in Nottinghamshire. With the possibility of an "energy crisis" coal-firing will probably continue to compete effectively with oil-fired and atomic power stations.

11: *Conclusion*

The reciprocating steam engine, if not quite dead, is certainly being used less and less in industry and the public services. As we have seen, few of them are still in use. But the age of steam is far from dead. In the form of the steam turbine coupled to electricity generators, steam power remains the method of creating most of the energy consumed in modern industrial societies. Furthermore, with a world shortage of oil fuels looming ahead, the internal combustion engine will probably be unable to continue to challenge steam power. So we may look forward to a revival of steam power, or at least steam generated electricity, on the roads and railways. However, there is as yet no prospect of a convenient alternative to the gas turbine for air transport.

Steam power, in short, has a bright future. Since the internal combustion engine is dependent on oil fuels and natural gas, reserves of which are running out; and since the problem of the *direct* conversion of atomic energy into electricity has not been solved, steam power is still the most reliable, efficient and flexible source of power available to modern societies. It seems certain, therefore, that steam power will be with us for some time to come and that it will continue to make a substantial contribution to industrial production and domestic comfort. It can be made available easily over long distances by high-voltage transmission cables so that all parts of the world can take advantage of it. The period of unchallenged supremacy which it enjoyed for much of the nineteenth century has long since disappeared. But man's ability to harness the elastic properties of steam remains one of his most useful skills. There is no reason to think that it will not remain important well into the twenty-first century and beyond.

In conclusion, we can say that the steam engine was developed to meet a succession of different needs which appeared in the eighteenth century as Britain became more industrialized. It was developed further as industrialization spread to the rest of the world in the nineteenth century. Once its tremendous value to society was recognized, a steady stream of capital, skilled labour and other resources was put into its development. In both the eighteenth and nineteenth centuries, there was great scope for it to fulfil existing needs and to create new needs. Indeed, the steam engine was responsible for the industrialization of Western Europe and the rise in its standard of living in the last two centuries.

We have traced the main lines of this remarkable achievement in this book. But there have always been people who criticize material progress, and because the steam engine has helped such progress, it, too, has been criticized. Humanitarians, at the time when child labour in the factories was arousing concern, blamed the unrelenting rhythm of the steam engine for the human suffering. Samuel Butler (1835–1902) in his brilliant satire of 1872, *Erewhon,* wrote about a society which only escaped from being mastered by its "vapour engines" by scrapping all its machinery. But the steam engine has generally not been used as a means of repression. Admittedly, the steam warship made a significant impact on naval warfare, and steam transport on the railways has played a vital part in the manoeuvring of modern armies. Nevertheless, compared with the invention of the internal combustion engine or atomic power—or even the development of iron and steel technology—the steam engine has not been an instrument of war. It has mainly been used for pumping, winding and turning the wheels of industry and transport. More than most products of human technological ingenuity, the steam engine has benefited, and continues to benefit, mankind.

Considering how much we owe to the steam engine, it is to be hoped that the surviving monuments of the past achievements of the Steam Age will be preserved with respect. The industrial archaeology of steam power is

Below. An engraving by the artist, Gustav Doré, of the squalor in London caused by urban life, factories and the railways—all of which were dependent on the steam engine.

currently attracting a lot of attention in Britain and elsewhere. With so many reciprocating steam engines now obsolete or about to become obsolete, it is a matter of urgency that they should be preserved properly. Many engines, both stationary and locomotive, are already in museums of various kinds. But a truly representative selection of steam engines, covering the whole range of types and applications, must be conserved in order to do justice to their enormous contribution to Western civilization.

Above. The Iron Maiden, an old traction engine which was mainly used to provide power at fairgrounds. Hopefully, such engines will continue to be preserved as a memorial to the Steam Age.

Date Chart

*c.*AD 100	Hero of Alexandria designed the "aeolipile"
1644	Torricelli demonstrated that the atmosphere has weight
1662	Foundation of the Royal Society
1663	The Marquis of Worcester devised a pump worked by steam pressure
1672	Otto von Guericke experimented with a vacuum and air pressure
1690	Denis Papin condensed steam in a cylinder to produce a vacuum
1698	Thomas Savery's patent for a pump worked by "the Impellent Force of Fire"
1709	Iron ore first successfully smelted with coke as the fuel by Abraham Darby at Coalbrookdale
1712	First Newcomen engine erected to pump water from a coal mine
1769	James Watt's patent for the separate condenser Nicholas Cugnot's steam gun-carriage
1781–7	Improvements to steam engine patented by James Watt: sun and planet gearing for rotative action (1781); double action (1783); parallel motion (1784); steam governor (1787)
1800	Expiry of Watt's patents
1801	Richard Trevithick's high pressure engine installed in a vehicle
1802	William Symington's paddle tug *Charlotte Dundas*
1803	Arthur Woolf's patent for compound engine
1804	Trevithick's first railway locomotive ran on the Penydarren tramway Oliver Evans' high pressure engine in the U.S.A.
1807	Robert Fulton's paddle steamer *Clermont*
1825	Stockton and Darlington Railway opened
1830	Liverpool and Manchester Railway opened
1831	Sir Goldsworthy Gurney's steam carriage
1837	S.S. *Great Western* launched in Bristol

1838	*Sirius* and *Great Western* both crossed the Atlantic westwards under steam
1843	S.S. *Great Britain* launched in Bristol
1845	William McNaught patented method of compounding a beam engine
1849	G. H. Corliss invented the rapid acting steam inlet valves in the U.S.A.
1851	The Great Exhibition
1855	Sir Joseph Bazalgette appointed engineer to the Metropolitan Board of Works to construct a main drainage system for London
1859	S.S. *Great Eastern* launched in London
1865	The "Red Flag Act"
1866	Aveling and Porter's steam road roller
1869	Suez Canal opened
1874	P. W. Willans patent for high speed enclosed engines
1884	Charles Parsons patented his first successful reaction steam turbine
	Willans introduced his central valve high speed engine
1885	Todd's "uniflow" patent
1888	First Parsons turbo-alternator installed
1889	C. G. P. de Laval's patent for an impulse-type steam turbine
1890	Patent for enclosing engine and lubrication under pressure by Payne and Belliss
1896	Repeal of the "Red Flag Act" encouraged internal combustion engine in road vehicles.
1897	Spithead Naval Review: Parsons' turbine-powered ship *Turbinia* achieved 34 knots
1903	Wright Brothers achieve powered flight
1906	S.S. *Lusitania* and *Mauretania*—Cunard turbine-powered liners launched
1912	S.S. *Titanic,* White Star Liner, sank on her maiden voyage
1914–18	First World War
1926	National Grid introduced in Britain
1938	L.N.E.R. locomotive *Mallard* achieved record 126 m.p.h. for steam power
1970	Drax and other modern power stations installed with turbines of 600,000 kw. capacity

Acknowledgements

The authors and publishers would like to thank the following for permission to reproduce copyright illustrations on the pages indicated: B. J. Finch, *frontispiece*, 38, 67, 68, 89; British Rail, 32, 47 (*above*), 49, London Transport Executive, 52; Brunel University, 60, 63 (*above*); Mary Evans Picture Library, 6, 7, 14, 16, 72 (*below*); National Coal Board, 31, 74; North Thames Gas, 79 (*below*); G. Watkins, 34, 35, 72 (*above*), 73, 78 (*below*), 81; The Science Museum, 11, 12, 13, 15, 17, 18, 19, 28, 29 (*above*), 44, 47 (*below*), 48, 50, 56 (*opposite, below*), 58, 59, 62, 64, 83, 84; R. A. Buchanan, 26, 41, 66 (*above*), 88; Central Electricity Generating Board, 86; Sotheby & Co., *jacket picture*.

The publishers would like to thank Elizabeth Mackintosh for the drawings on pages 20, 22, 24, 36.

The authors would like to thank all those friends and colleagues who have helped them over the years in their study of the history of steam power. In particular they are grateful to Miss Judith Beaven for her assistance in producing the typescript for this book.

Further Reading

Baker, W. A., *From Paddle Steamer to Nuclear Ship* (Watts, 1965).

Barton, D. B., *The Cornish Beam Engine* (Barton, 1965).

Bonnet, H., *The Saga of the Steam Plough,* (David and Charles, 1971).

Buchanan, R. A., *Industrial Archaeology in Britain* (Penguin, 1972).

—, *Technology and Social Progress* (Pergamon, 1965).

Cardwell, D. S. L., *Technology, Science and History,* (Heinemann Educational, 1972).

Clark, R. H., *The Development of the English Steam Wagon* (Goose, 1963).

Davidson, C. B., *Steam Road Vehicles—A Historical Review* (Science Museum, 1970).

Derry, T. K. and Williams, T. I., *A Short History of Technology* (O.U.P., 1960).

Dickinson, H. W., *A Short History of the Steam Engine* (F. Cass, 1938, reprinted 1963).

—, *James Watt—Craftsman and Engineer* (Cambridge U.P., 1935).

— and Titley, A., *Richard Trevithick—the Engineer and the Man* (Cambridge U.P., 1934).

Ferneyhough, F., *Railways* (Wayland, 1970).

Harris, T. R., *Arthur Woolf—The Cornish Engineer* (Barton, 1970).

Hills, R. L., *Power in the Industrial Revolution* (Manchester U.P., 1970).

Hobday, P., *Man the Industralist* (Priory Press, 1973).

Law, R. J., *The Steam Engine—A Short Introduction* (H.M.S.O., 1965).

Musson, A. E., and Robinson, Eric, *Science and Industry in the Industrial Revolution* (Manchester U.P., 1969).

Norris, W., *Modern Steam Wagons* (David and Charles, 1906, reprinted 1971).

Parsons, R. H., *The Early Days of the Power Station Industry* (Cambridge U.P., 1940).

Pursell, Carroll W. Jr., *Early Stationary Steam Engines in America* (David and Charles, 1969).

Robinson, Eric, and Musson, A. E., *James Watt and the Steam Revolution* (Adams and Dart, 1969).

Rolt, L. T. C., *Thomas Newcomen—The Prehistory of the Steam Engine* (David and Charles, 1963).

—, *I. K. Brunel* (Longman, 1957).

—, *George and Robert Stephenson* (Longman, 1960).

Rowland, R. T., *Steam at Sea* (David and Charles, 1970).

Smith, E. G., *A Short History of Naval and Marine Engineering* (Cambridge U.P., 1938).

Storer, J. D., *A Simple History of the Steam Engine* (J. Baker, 1969).

Watkins, G., *The Stationary Steam Engine* (David and Charles, 1968).

—, *The Textile Mill Engine,* Vols 1 & 2, (David and Charles, 1970 and 1971).

— and Buchanan, R. A., *Industrial Archaeology of the Stationary Steam Engine* (publication pending).

Glossary

AIR PUMP[removed the condensed water and air contained in the steam; and was usually driven from the engine itself.

ANNULAR CYLINDERS: a concentric arrangement with a high pressure cylinder inside a low pressure cylinder.

BEAM ENGINES: transmitted the effort from the piston rod to the crank by a massive lever or beam usually of cast iron. In some cases the beam was placed at the bottom of the engine or the sides (i.e. the *side lever*) and in other cases the beam was suspended from one end, with the cylinder at the other end, and the connecting rod between (i.e. the *grasshopper beam*).

COMPOUND ENGINES: exhausted the steam into two, three, or four cylinders of increasing volume in succession expanding it all the time.

CONDENSERS: were airtight chambers into which the exhaust steam passed for cooling back to water after leaving the engine. Cooling was by a jet of cold water, or the cold water was passed through numerous tubes to condense the steam outside them.

CRANKS: were rotating levers attached to the crankshaft which were needed to convert the thrust of the piston rod into a rotary motion for driving.

CYLINDER: a casting of very hard iron, bored accurately to allow the piston to move freely from end to end without leakage. It was closed at each end by covers, one or both of which were fitted with glands, allowing the piston rod to move very freely. It was provided with ports to admit and exhaust the steam, drains to release condensed water and feet for attaching it to the engine bed.

DYNAMO: a machine which produces an electric current.

FEED PUMP: usually a simple pump which returned the condensed feed water to the boiler.

FLYWHEEL: a large heavy wheel fixed on the crankshaft.

GLANDS: recessed bosses in the cylinder cover and valve chests which were fitted with packings of fibre or metal segments, which allowed the moving rods to work freely.

GOVERNOR: consisted of rotating weights which varied the steam supply to the cylinder if the speed was too high or low.

GUIDES: were necessary to resist the side thrust from the connecting rod as it delivered the effort to the crank.

OSCILLATING ENGINES: were the type in which the cylinder was not solidly fixed to the engine frame. The piston rod was connected directly to the crank pin, so that the cylinder had to swing or oscillate to follow the movement of the crank pin.

PISTON: a disc turned to fit the cylinder bore, with a hole in the centre bored to fit the end of the piston rod.

PISTON ROD: a circular rod attached to the piston by which the effort was transmitted to the crosshead. It was continued through the rear cylinder cover, if a supporting guide was fitted at that end, or if there were two cylinders arranged in tandem.

RECIPROCATING: an alternating backward and forward movement.

RETORTS: an enclosed vessel in which coal is heated to extract gas.

TANDEM ENGINES: horizontal engines with the cylinders placed one behind the other.

THE McNAUGHT COMPOUND BEAM ENGINE: had the high pressure cylinder taking the steam from the boiler placed between the centre of the beam and the connecting rod, with the low pressure cylinder at the other end.

THE WOOLF COMPOUND BEAM ENGINE: had the two cylinders placed together, the steam again passing from the smaller high pressure to the low pressure cylinder and the condenser.

THERMODYNAMICS: the science of the relationship between heat and work done.

TRUNNIONS: the pivots on which a swinging cylinder is suspended.

UNIFLOW ENGINE: an engine in which steam enters the cylinder at one end and exits at the other, rather than reversing its direction as in other engines.

VALVES: were fitted with flat or circular faces, which moved upon the ports in the cylinder to admit and exhaust the steam. Slide and piston valves moved to and fro. The Corliss valves oscillated over the port faces. Drop valves were simply lifted and lowered upon the seats. They were operated by an eccentric or cam device with a very short movement. A separate manually-operated stop valve was also necessary to stop the engine, although the pressure was maintained in the piping from the boiler.

VERTICAL ENGINES: had the working parts in a vertical centre line, with the cylinder at the top or the bottom of the engine.